CHAOS AND DYNAMICAL

SYSTEMS

CHAOS AND DYNAMICAL SYSTEMS

David P. Feldman

PRINCETON UNIVERSITY PRESS

Princeton & Oxford

Published by Princeton University Press
41 William Street, Princeton, New Jersey 08540
6 Oxford Street, Woodstock, Oxfordshire OX20 1TR

press.princeton.edu

All Rights Reserved

LCCN: 2019936804
ISBN 978-0-691-16152-5

British Library Cataloging-in-Publication Data is available

Editorial: Vickie Kearn, Susannah Shoemaker, and Lauren Bucca
Production Editorial: Nathan Carr
Text Design: Pamela L. Schnitter
Jacket/Cover Design: Pamela L. Schnitter
Jacket/Cover Credit: April Vollmer, *Kosha 9*, 2010, woodcut and
mixed media collage on panel, 12 × 12 inches.
Production: Erin Suydam
Publicity: Matthew Taylor and Kathryn Stevens
Copyeditor: Lor Campbell Gehret

This book has been composed in Adobe Garamond
and Helvetica Neue

Printed on acid-free paper. ∞

Printed in the United States of America

1   3   5   7   9   10   8   6   4   2

CONTENTS

PREFACE

The goal of this book is to give a conceptual tour of the fields of dynamical systems and chaos. What are the important themes and realizations that have emerged from these areas of study? What are the lessons of dynamical systems that are of particular importance for the study of complex systems? My aim is to give you a firm understanding of key concepts and ideas, including chaos, the butterfly effect (known scientifically as sensitive dependence on initial conditions), bifurcations, and strange attractors.

A *dynamical system* is any mathematical system that changes in time according to a well specified rule. There are two types of dynamical systems that I will discuss in this book: differential equations and iterated functions. Differential equations play a central role in the physical sciences, and increasingly in the natural and social sciences as well. Iterated functions, in which a function is repeatedly applied to a number or a vector, are less common in the sciences, but mathematicians have used them extensively to study properties of dynamical systems.

Dynamical systems also refers to an area of applied mathematics. This area includes the study of particular scientific applications of dynamical systems, but also takes a broader and more general look at dynamical systems, developing methods for analyzing and characterizing dynamical behaviors and seeking to understand the range of behaviors exhibited by different classes of models. These broader concerns are the focus of this short book.

The field of dynamical systems is sometimes referred to as "chaos theory," although this term tends to make many who study dynamical systems, including me, a bit uneasy. There has been so much hype around chaos theory—some of it deserved, some of it not so much—that there is a tendency for scientists and mathematicians to avoid the term. Also, chaos refers to a particular type of behavior, while the study of dynamical systems includes a broader range of phenomena than just chaos, and so the term "chaos theory" may be misleadingly narrow. Additionally, "chaos theory" is not really a theory in the sense that the term theory is typically used: a broad, general explanatory framework. Chaos is a fascinating and fun phenomenon seen in mathematical and physical systems, but it is not an explanatory framework in the same way as, say, electromagnetic theory, plate tectonics, or evolution by natural selection.

This book is not intended to be a comprehensive text on dynamical systems. Rather, I hope that it fills a gap between two types of book. On the one hand, there are some excellent and engaging popular books on chaos and dynamical systems intended for the lay reader, such as Gleick (1987) and Stewart (2002). There are also outstanding textbooks on dynamics aimed at math or physics majors, or those with a sturdy math background (e.g., Peitgen et al. (1992), Kaplan and Glass (1995), Strogatz (2001), and Hilborn (2002)). This primer aims to serve as bridge between these two types of books. I want to communicate the important ideas from dynamical systems using just enough math to put flesh on the bones, but not go into as much detail, or have as high a mathematical hurdle for entry, as is expected by most texts. For those who want to dig deeper, suggestions for further reading can be found at the end of each chapter.

I've chosen to focus on aspects of chaos and dynamical systems that I think are particularly relevant to the study of complex systems. There is not agreement about just what the field of complex systems is, but I think there is some consensus that complex systems usually involves the study of systems with many components which interact and are often heterogeneous. Complex systems often have properties that are not explicitly contained in or easily derived from a knowledge of the system's constituent parts and their interactions. Such properties are said to be *emergent*. I'll say some more about complex systems and emergence in Sections 10.2 and 10.3.

A bit about the sequence of topics in this book, whose dependencies are illustrated in Fig. 0.1. Chapters 1 and 2 introduce the two types of dynamical systems that we will examine in this book: iterated functions and differential equations, respectively. Chapter 3 is an interlude in which I discuss some of the unspoken assumptions behind Newtonian mechanics and, implicitly, much of science. This chapter also includes observations on the different ways that models are used in the sciences. Chapter 3 refers only in a general way to the dynamical systems from Chapters 1 and 2, and so I think Chapter 3 could be read first if you want to start on a more philosophical note.

In Chapter 4 I introduce chaos, including sensitive dependence on initial conditions. This chapter uses only iterated functions, so it is possible to read Chapter 4 without having read Chapter 2 if you're so inclined. The discussion of chaos continues in Chapter 5, where I dig into some broader questions about the implications of the butterfly effect and different ways of thinking about and defining randomness.

In Chapter 6 I look at the phenomenon of bifurcations: sudden, qualitative changes in a system's behavior. This chapter makes use of differential equations, and so depends only on Chapter 2. Chapter 7 is about universality in chaos, and requires only Chapter 4. In Chapter 8 I introduce phase space and higher-dimensional dynamical systems, and in Chapter 9 I consider strange attractors as well as attractor reconstruction and the Lorenz map and Poincaré sections. These two chapters (8 and 9) draw on materials from all previous chapters except for 6 and 7.

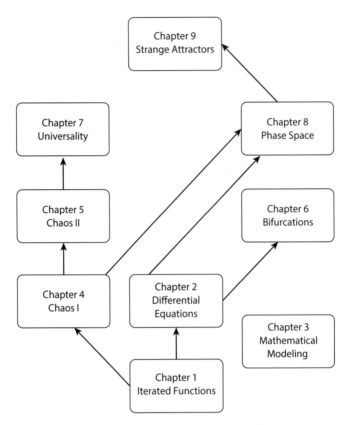

Figure 0.1. Dependencies among the chapters of this book.

In Chapter 9 I briefly discuss phase-space reconstruction and delay coordinates, techniques from non-linear time series analysis. Finally, I end in Chapter 10 with some concluding thoughts.

I have included some topics that aren't traditionally part of dynamical systems books, but which are encountered frequently in the study of complex systems. Examples include emergence in Section 7.3 and power laws, phase transitions, and related notions in Section 7.8. I've also omitted some standard dynamical systems topics that I don't think have much relevance to complex systems;

most significantly, Hamiltonian systems. There are some topics that are important for the study of complex systems—such as synchronization, pattern formation, and information theory—that I would have liked to include, but decided to omit to keep the length of this book reasonable.

What math background is needed for this book? I assume that readers are comfortable with basic algebra and functions and are familiar with the basic notions of differential calculus. Specifically, readers should understand the idea of the derivative as a function's instantaneous rate of change. But to read this book you will not need to use the techniques of calculus (the chain rule, product rule, and so on) to calculate derivatives, nor will I use integration. In practice, I have found that almost anyone with a moderate affinity for mathematics can learn this material with a bit of work. A class in differential calculus would be helpful background, but is definitely not necessary.

If you have a more extensive math background, I think you will still get a lot out of this book. I will be presenting material that you may not have seen before, even if you have taken a class in differential equations. In my experience students with a wide range of math backgrounds find chaos and dynamical systems to be accessible and engaging. Although I avoid most advanced math, I have tried my best to avoid dumbing down the material.

I have taught a massive, open, online course (MOOC) on dynamical systems and chaos that is structured similarly to this book and is at a similar level. The MOOC is offered regularly as part of the Santa Fe Institute's Complexity Explorer Project. The current version of the course can be found at http://chaos .complexityexplorer.org. Here you can find over 14 hours of videos I've produced on chaos and dynamical systems. Site registration and all course materials are free.

If you encounter any errors in the book, I would be grateful to learn of them. I hope you enjoy reading this book and learning about the fun world of dynamical systems.

Acknowledgments

I thank John Miller for encouraging me to write this book and being patient as it slowly came to fruition. Comments from anonymous reviewers were thoughtful and helped me improve the manuscript. My understanding and appreciation of dynamical systems has benefited over the years from numerous discussions with colleagues, including Liz Bradley, Aaron Clauset, Jim Crutchfield, Joshua Garland, David Krakauer, Richard Levins, Melanie Mitchell, and Cosma Shalizi. Rebecca Wood provided technical issues with several of the figures. Developing a MOOC on chaos and dynamics was challenging but very rewarding. I thank Melanie Mitchell and the Santa Fe Institute's Complexity Explorer team for inviting me to contribute a course and for their help and encouragement. And I thank Complexity Explorer students from all over the world who took my classes and participated in online discussions.

I am grateful to Vickie Kearn and the team at Princeton University Press, including Nathan Carr and Susannah Shoemaker. I am especially appreciative of the expert copy-editing performed by Lor Gehret. Of course I take full responsbility for any errors that escaped Lor's watchful eyes.

I am grateful for musical inspiration from: Aly & Fila, Andski, Gai Barone, Cosmic Gate, Gareth Emery, Gabriel & Dresden, Kaeno, Markus Schulz, Monoverse, Paul Oakenfold, Solarstone, and Ian Standerwick. In exchange for food, I have received ample feline inspiration from Ancho, Apple, and Panda. I think this is going to be my last book on chaos, and so this is my last chance to use a dynamical systems reference in an acknowledgment. So I thank Doreen Stabinsky for her love, companionship, inspiration, and for helping keep my largest Lyapunov exponent from getting too positive.

David Feldman
Mount Desert, Maine, USA

CHAOS AND DYNAMICAL

SYSTEMS

1

INTRODUCING ITERATED FUNCTIONS

A *dynamical system* is any mathematical system that changes in time according to a well specified rule. We will look at two different types of dynamical systems in this book: iterated functions and differential equations. We will use these two types of dynamical systems to address a central question: what sorts of behaviors are possible for different types of dynamical systems?

In this chapter I'll introduce iterated functions. I'll say what iterated functions are, present several ways of visualizing their behavior, and introduce some key terminology. This chapter may be a bit abstract. We'll approach iterated functions as simple mathematical systems, without attention to their roles as models of the physical or biological world. In Chapter 2, where I introduce differential equations, we will begin to see how dynamical systems are used in science. This chapter may also be a bit dry; there's nothing too deep in the next few pages. However, it's essential background for the more interesting and surprising results that will come later.

1.1 Iterated Functions

Consider a function f of one variable such as $f(x) = x^2$. A function establishes a relationship between a set of numbers, the inputs x, and another set of numbers, the outputs $f(x)$. We can think of

the function f as an action. We start with a number x, apply the
function f to it, and get a new number. This new number is called
$f(x)$: it is x after it has had f applied to it.

Usually we think of the application of a function as a one-
shot deal: Do f to x, get $f(x)$, end of story. For example, if
$f(x) = x^2$, then $f(3) = 3^2 = 9$, and $f(-0.5) = (-0.5)^2 = 0.25$.
But if we apply the function repeatedly, using the output at one
step as the input to the next, then we have a dynamical system:
a mathematical entity that changes in time according to a well-
defined rule. For example, we could start with 3, apply f, and
obtain 9. Apply f again, and we get $9^2 = 81$. Apply f yet again,
and we get $81^2 = 6561$. The result is a sequence of numbers:

$$3, 9, 81, 6561, 43046721, \ldots . \qquad (1.1)$$

We see that the numbers quite quickly become very large.

This process is known as *iteration*. The application of f is
repeated, or iterated, and the output of one step is used as the input
for the next step. Our starting number—in this case 3—is known
as the *seed* or the *initial condition*. The sequence of numbers in
Eq. (1.1) is known as the *orbit* or *itinerary* of 3. The initial condition is usually denoted x_0. The first value in the itinerary is denoted
x_1 and is called the first *iterate*. This value is obtained by applying
f to x_0. That is, $x_1 = f(x_0)$. The second iterate is denoted x_2 and
is obtained by applying f to x_1: $x_2 = f(x_1)$. Equivalently, we may
think of x_2 as resulting from twice applying f to x_0: $x_2 = f(f(x_0))$.
Subsequent iterates are denoted similarly.

Let's consider another example: $g(x) = \frac{1}{2}x + 4$. I'll choose an
initial condition of $x_0 = 1$. The first iterate x_1 results from g acting
on x_0:

$$x_1 = g(x_0) = g(1) = \frac{1}{2}(1) + 4 = 4.5 . \qquad (1.2)$$

Subsequent iterates are found in a similar manner; $x_2 = g(x_1)$, and
so on. The first several iterates of this initial condition are shown in

t	x_t
0	1
1	4.5
2	6.25
3	7.125
4	7.5625

Table 1.1 The first several iterates of the initial condition $x_0 = 1$ for the function $g(x) = \frac{1}{2}x + 4$.

Table 1.1. (You might want to grab a calculator and take a moment to verify these numbers.)

It is often useful to display an orbit graphically rather than in a table or list. The orbit in Table 1.1 is plotted in Fig. 1.1. This type of graph is known as a *time series plot*. Such a graph gives a clear view of the orbit's behavior. In Fig. 1.1 we can see that the orbit is approaching 8. Note that a time series plot is not a graph of the function that is being iterated, $g(x) = \frac{1}{2}x + 4$. Instead, it is a plot of the orbit or itinerary.

The value of 8 is a *fixed point* of the function $g(x)$. This means that 8 does not change if g operates on it: $g(8) = 8$, as we can verify:

$$g(8) = \frac{1}{2}8 + 4 = 4 + 4 = 8. \tag{1.3}$$

In general, a number x is a fixed point of $f(x)$ if it is a solution to the equation

$$f(x) = x. \tag{1.4}$$

Such an equation is called a *fixed-point equation*. In words, Eq. (1.4) says that x, when acted on by f, yields x. A function can have any number of fixed points, including none at all.

Iterated functions are our first example of a dynamical system, a mathematical system that changes in time according to a

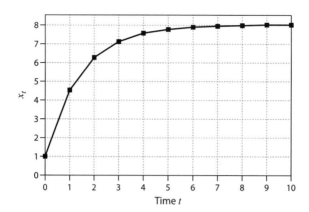

Figure 1.1. The time series plot for the initial condition $x_0 = 1$ iterated with $g(x) = \frac{1}{2}x + 4$.

well-defined rule. In the example we just considered, the rule is given by the function $g(x) = \frac{1}{2}x + 4$. The dynamical system that results from iterating this equation is sometimes written as:

$$x_{t+1} = \frac{1}{2}x_t + 4. \tag{1.5}$$

This notation makes the dynamical nature of the equation clearer. We can see that the next value of x is determined by the current value of x. That is, x_{t+1} is a function of x_t. Thus, as long as we know the initial value x_0, we can figure out all subsequent values of x by repeated application of Eq. (1.5).

Note that Eq. (1.5) does not directly tell us x_t as a function of t. If, say, we want to know x_{13}, we can't just plug in $t = 13$ somewhere. Rather, we need to start at some known value of x, usually x_0, and iterate forward, one step at a time, using Eq. (1.5). Doing so might be a bit time consuming, but it is at base a very simple procedure. There is a rule—namely the function $f(x)$— and that rule is applied again and again. For all but the simplest such functions one almost always turns to a computer to carry out the iterations. I used a short python program to iterate the function $g(x) = \frac{1}{2}x + 4$ and make the time series plot shown in Fig. 1.1.

Before going on I should mention some additional terminology. A function f takes input values and returns output values. So one sometimes refers to f as a *mapping* from input to output. Iterated functions are also often called *maps*. In mathematics, a map is synonymous with function. I will usually refer to functions as functions, but the term map is very commonly used in dynamical systems so you will likely see it elsewhere.

1.2 Thinking Globally

In the study of dynamical systems we are often interested not in the numerical values of a particular orbit, but in its long-term behavior. We want to know about the big picture—the global dynamics of the function—not the local details of each and every point in a particular orbit. For example, when describing the itinerary of $x_0 = 1$ when iterated with $g(x) = \frac{1}{2}x + 4$, we simply say that it approaches 8, rather than list all the data in Table 1.1.

Let's consider another example: the square root function $f(x) = \sqrt{x}$. Our goal will be to figure out the long-term behavior of *all* initial conditions. (Since the square root of a negative number results in a complex number—also known as an imaginary number—we will limit our analysis to non-negative numbers.) To get us started, let's choose the seed $x_0 = 4$. Then x_1 is the result of applying the function to x_0. Since $f(4) = \sqrt{4} = 2$, the first iterate is 2. The next iterate is approximately 1.414, since $\sqrt{2} \approx 1.414$. We keep on square rooting and obtain the itinerary shown in Table 1.2.

The orbit for $x_0 = 4$ is shown in Fig. 1.2. Also on this figure are the time series plots for three other initial conditions: 2, 0.5, and 0.25. You can obtain orbits for these initial conditions by entering the seed and then repeatedly hitting the $\sqrt{}$ key on your calculator. However, without using a calculator we can understand the shape of the time series plots qualitatively.

t	x_t
0	4
1	2
2	1.414
3	1.189
4	1.091
5	1.044

Table 1.2 The first several iterates of the initial condition $x_0 = 4$ for the function $f(x) = \sqrt{x}$.

When you take the square root of a number larger than one, the result is a smaller number. For example, $\sqrt{10000} = 100$, $\sqrt{16} = 4$, and $\sqrt{1.5} \approx 1.225$. Numbers larger than 1 get closer and closer to 1 when successively square rooted. We can see this in Fig. 1.2. The seeds 4.0 and 2.0 are both getting smaller and approaching 1.

On the other hand, numbers between 0 and 1 get *larger* when square rooted. For example $\sqrt{0.25} = 0.5$. It might be easier to see this using fractions:

$$\sqrt{\frac{1}{4}} = \frac{1}{2}, \text{ because } \left(\frac{1}{2}\right)^2 = \left(\frac{1}{2}\right)\left(\frac{1}{2}\right) = \frac{1}{4}. \tag{1.6}$$

So for this dynamical system—iterated square rooting—any number between 0 and 1 will increase and approach 1, and any number larger than 1 will decrease and approach 1. The numbers 0 and 1 are fixed points; they are unchanged when square rooted: $\sqrt{0} = 0$, and $\sqrt{1} = 1$.

With these observations, we can describe the global dynamics of the square root function. That is, we can specify the long-term behavior of all non-negative initial conditions. Any initial condition larger than 1 will get smaller and move closer and closer to 1. Any initial condition between 0 and 1 will get larger and get closer

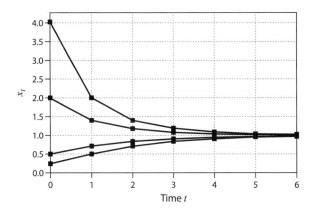

Figure 1.2. The time series plot for four different initial conditions iterated with $f(x) = \sqrt{x}$.

and closer to 1. The initial conditions 0 and 1 are fixed points. They do not change when acted upon by the function: $\sqrt{0} = 0$ and $\sqrt{1} = 1$.

1.3 Stability: Attractors and Repellors

The fixed points of $f(x) = \sqrt{x}$ are 0 and 1, but these two fixed points have a rather different character. The fixed point at 1 is called *stable* or *attracting*. Nearby orbits are pulled toward it; it attracts nearby points. It is called stable because if one is at the fixed point and then a perturbation moves you a little bit away, you will return to the fixed point. That is, if you are at 1 and something happens and you get moved to 1.1, the square-rooting function will move you back, closer and closer to 1. The first several iterates of 1.1 are: 1.1, 1.049, 1.024, 1.012. The orbit is getting closer to 1.

The fixed point at 0 is different. You will not be surprised to learn that this fixed point is *unstable* or *repelling*. If you are at 0 and something happens and you get bumped to 0.05, you will not return to 0. Instead, you will get pushed away from 0, never

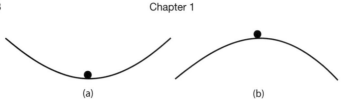

Figure 1.3. A schematic illustration of (a) a stable and (b) an unstable fixed point.

to return. The first several iterates of 0.05 are: 0.05, 0.224, 0.473, 0.688. The orbit is not getting closer to 0.

Stable and unstable fixed points are illustrated schematically in Fig. 1.3. On the left is shown a stable fixed point—a ball at the bottom of a valley. If the ball is moved a small amount it will return to the bottom of the valley. On the right, the fixed point is unstable. If the ball is moved a little bit it will roll down one side of the hill, not to return.

For completeness, I should mention that it is possible for a fixed point to be poised between stability and unstability. In this case, if one moves away from the fixed point one neither returns to the fixed point nor is pushed away. Fixed points with these properties are called *neutral*. In terms of the schematic representation of fixed points shown in Fig. 1.3, neutral fixed points look like a ball resting on a perfectly flat table. If the ball is moved to the left or right, it will stay there; it won't return to its original location, but it also won't roll further away.

The stability of fixed points—or of other dynamical behavior that we will encounter later—is an important notion. Typically, in a mathematical model or the real world, one only expects to observe stable fixed points. An unstable fixed point is susceptible to a small perturbation; a tiny external influence will move the system away from the unstable fixed point. For example, it is possible to carefully balance a pencil on its eraser. However, it will not stay this way for long. A small vibration or bit of wind will make the pencil fall over and lie on its side. Or, returning to Fig. 1.3,

Figure 1.4. A stable fixed point. A large perturbation and the ball will not return to the fixed point, but for small perturbations it will return.

we would not expect to observe the situation depicted in part (b). A rock balanced on the top of a hill will not remain there indefinitely. A small gust of wind or a little push will cause it to roll downhill. The upshot is that in dynamical systems one is particularly focused on stable behavior, as usually it is only stable behavior that is observed.

Before moving on, I should define stable and unstable fixed points just a bit more carefully. A fixed point x is stable if there is an open interval around x such that any initial conditions in this interval get closer and closer to the fixed point x. In terms of the schematic view of Fig. 1.3, this says that a fixed point is a point in the bottom of a valley, regardless of how wide or narrow the valley is. This is illustrated in Fig. 1.4.

1.4 Another Example

Let's consider another example: the cubing function $f(x) = x^3$. What are its dynamics? How many fixed points are there and what are their stability? Let's start by solving for the fixed points. A point x is fixed if it is unchanged by the function. That is, $f(x) = x$. Here, the function is $f(x) = x^3$, so the equation for fixed points is:

$$x^3 = x. \tag{1.7}$$

This equation has three solutions: $-1, 0$, and 1. Each of these numbers, when cubed, does not change. For example, $(-1)^3 = (-1)(-1)(-1) = -1$.

Are these fixed points stable or unstable? Let's think about what happens to different initial conditions. A number larger than 1 will get larger when cubed. For example, the itinerary of the initial condition $x_0 = 2$ is:

$$2, 8, 512, 134217728, \ldots . \tag{1.8}$$

The orbit grows very rapidly and will keep getting larger. One describes this situation by saying that the orbit grows without bound or tends toward infinity. A number less than -1 will get "bigger and more negative." (Strictly speaking this means that the orbit gets smaller; as one moves to the left on a number line the numbers get smaller. Negative three is less than negative two.) So we say that the orbit of -1 decreases without bound or tends toward negative infinity. Lastly, numbers between -1 and 1 will approach zero when cubed. For example, the orbit of $x_0 = -0.9$ is:

$$-0.9, -0.729, -0.38742, -0.058450, \ldots . \tag{1.9}$$

Thus, 0 is a stable, or attracting fixed point; it pulls in all initial conditions between -1 and 1. The fixed points at ± 1 are unstable, or repelling.

1.5 One More Example

I'll end this chapter with one more example. We'll consider the function $f(x) = x^2 - 1$. Does this function have any fixed points? Yes—two of them, in fact. The fixed point equation

$$f(x) = x \tag{1.10}$$

has two solutions:

$$x = \frac{1}{2}(1 + \sqrt{5}) \approx 1.61803 , \tag{1.11}$$

and

$$x = \frac{1}{2}(1 - \sqrt{5}) \approx -0.61803 . \tag{1.12}$$

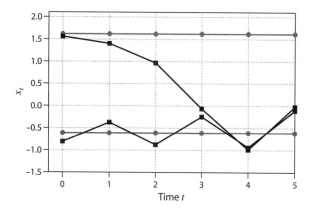

Figure 1.5. The time series plot for four different initial conditions iterated with $f(x) = x^2 - 1$. The two orbits plotted with gray circles are the fixed points, 1.618 and -0.618.

Are these fixed points stable? Let's iterate and see. In Fig. 1.5 I have plotted the orbits for four different initial conditions. The orbits shown with gray circles are the two fixed points, $x \approx 1.61803$ and $x \approx -0.61803$. The orbits plotted with squares begin close to the fixed points; the top orbit has an initial condition of $x_0 = 1.55$, and the initial condition for the bottom orbit is $x_0 = -0.8$. We see in the figure that the two square orbits are not pulled in toward the fixed points, so the fixed points are not stable.

It looks like the two orbits plotted with squares are getting closer together. By $t = 4$ or 5 they are almost on top of each other. What could be going on? To address this question, in Fig 1.6 I have plotted the two square orbits out to $t = 15$. The orbits of the two fixed points are again shown as gray circles. As in the previous figure, we see that the two square orbits do not get pulled toward the fixed points. Instead, the two orbits both approach periodic behavior; they oscillate between -1 and 0. These two points, -1 and 0, form a cycle of period 2.

To see that the orbit of -1 is periodic, first, we let f act on -1:

$$f(-1) = (-1)^2 - 1 = 1 - 1 = 0 \,. \tag{1.13}$$

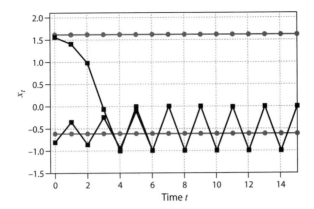

Figure 1.6. The time series plot for the four different initial conditions iterated with $f(x) = x^2 - 1$. The two orbits plotted with grey circles are the fixed points, 1.618 and -0.618.

Then, we let f act on 0:

$$f(0) = 0^2 - 1 = -1. \qquad (1.14)$$

Thus, -1 is periodic with period two. The period is two, because it takes two iterations to return to the initial condition.[1] In other words, $f(f(-1)) = -1$. It thus follows, of course, that 0 is also periodic with period two.

The period-two cycle is attracting or stable. Nearby orbits are pulled in to the cycle. Figure 1.7 gives us another way to see this. In this figure I have made time series plots for 200 different initial conditions distributed uniformly between -1.6 and 1.6. One can see in the figure that all initial conditions get pulled quite quickly into the period-two attractor. Not all of the orbits are in phase.

---

1. The initial condition -1 is also periodic with period *four*, because -1 will return to itself after four iterations. The period of a periodic point is quite sensibly defined to be the smallest number of iterations needed for the point to return to itself. (In more formal mathematical settings, it is common to use the term *prime period* to refer to the smallest number of iterations needed for a point to return to itself. Then one would say that -1 is periodic with period two, four, six, and so on, but that it has a prime period of two.)

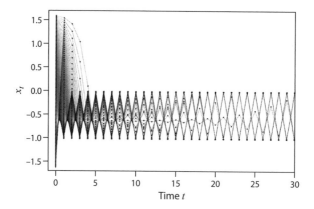

Figure 1.7. The time series plot for 200 initial conditions for the function $f(x) = x^2 - 1$. All initial conditions are pulled toward the period-two cycle at -1 and 0.

At, say, $t = 35$, about half of the orbits will be at (or very near to) -1 and half at 0. But all initial conditions get pulled in to the cycle. Said another way, the long-term behavior of all initial conditions between -1.6 and 1.6 is period two.

Actually, this is not quite right. It is not strictly the case that *all* initial conditions become period-two. The two fixed points, $x \approx 1.61803$ and $x \approx -0.61803$ are not period-two; they are fixed and so remain constant. So I need to amend the statements at the end of the last paragraph. I should have said: *almost* all initial conditions between -1.6 and 1.6 are pulled toward a period-two attractor. The word "almost" in this sentence has a technical meaning: it means that there are infinitely many more points that are pulled toward the period-two attractor than are not. Another way to say this is that if I choose an initial condition at random between -1.6 and 1.6, with probability 1 the orbit will get pulled toward the period-two attractor. This points out again the importance of stability and instability. The two unstable fixed points do not appear at all in Fig. 1.7. In order to observe the unstable behavior I would have to choose an initial condition *exactly* on the fixed point, something that is vanishingly improbable.[2]

To summarize, the function $f(x) = x^2 - 1$ has an attracting cycle of period two: -1 and 0. Equivalently, one says that the cycle is stable. If an orbit is cycling between -1 and 0 and then is perturbed slightly, it will return to the cycle. The function has two fixed points, but they are unstable, and thus do not affect the long-run behavior of almost all initial conditions.

1.6 Determinism

Before concluding, I have a few initial thoughts on an important concept and a recurring theme in the study of dynamical systems: *determinism.* The iterated functions we have studied in this chapter are all examples of *deterministic* dynamical systems. This means that there is no element of chance in the rule. For such dynamical systems the current value of x determines the next value, that value of x then determines the next value, and so on.

Thus, if one knows the function and the initial condition, then the entire future—that is, the itinerary—follows. One might think that deterministic systems are rather dull; once one writes down the rule and specifies the initial condition the story is essentially over. But one of the central lessons of dynamical systems is that deterministic systems still hold plenty of surprises. In Chapter 3 I will make some more extensive remarks on determinism and

---

2. There is a bit more mathematical fine print. It could also be the case that I chose an initial condition that after a finite number of iterations lands exactly on the unstable fixed point. This is also exceedingly unlikely; it occurs with probability zero. There are a countably infinite number of initial conditions that eventually land exactly on one of the unstable fixed points, but there are an uncountably infinite number of points on the interval between -1.6 and 1.6. Thus, there is zero probability that an orbit lands exactly on one of the fixed points after a finite number of iterations. The bottom line is that we do not expect to observe the unstable fixed points.

related issues. And then in later chapters we will encounter examples of deterministic dynamical systems that behave in ways that are counter-intuitive and produce results that are, in a sense, random.

1.7 Summary

A dynamical system is a mathematical system that changes in time according to a well-specified rule. In this chapter I introduced a simple type of dynamical system: iterated functions. Iterating a function is a repetitious and simple-minded task, requiring only a calculator and a bit of patience. One just applies a rule—in this case a function—to an initial condition over and over and over again.[3] Typically we're interested in a global view of the dynamics. How many fixed points does the dynamical system have and what are their stabilities? What is the long-term behavior of almost all initial conditions? In the examples in this chapter we have seen several types of long-term behavior. An orbit can tend toward positive or negative infinity, or get pulled to an attracting fixed point or an attracting cycle. In Chapter 4 we will see that iterated functions are capable of other, much more complex behavior.

To be honest, I hope this chapter was almost boring. My aim was to introduce a very simple type of dynamical system and to present some key terminology and concepts: initial condition or seed, orbit or itinerary, fixed points, and stable/unstable

---

3. I have presented the study of iterated functions as an experimental endeavor: choose a seed, grab your calculator, iterate, and see what happens. This is the approach that I'll take in this book. However, there are analytic and less computational approaches to studying the properties of iterated functions. See, e.g., Devaney (1989); Peitgen et al. (1992). These analytic techniques are a lot of fun and are a useful and important complement to the experimental approach I take here.

or attracting/repelling behavior. There is nothing deep or profound in this chapter. We will soon see, however, that simple iterated functions similar to the ones introduced here are capable of surprising—and definitely not boring—behavior. Before doing so, in the next chapter, I will introduce another type of dynamical system: differential equations.

2

INTRODUCING DIFFERENTIAL
EQUATIONS

A dynamical system is a mathematical system that changes in time according to a well specified rule. In the previous chapter I introduced iterated functions, our first example of a dynamical system: The rule in that case was simply a function $f(x)$ that is applied repeatedly. In this chapter I will introduce differential equations, another type of dynamical system. I will do so largely via an example—the temperature of a container of cold water. Differential equations are a bit more involved mathematically than iterated functions, so this chapter is a bit longer than the previous one.

2.1 Newton's Law of Cooling

I am writing this on a moderately hot day in July. On my desk I have a container of cold water (see Fig. 2.1) that I got not too long ago from the water cooler down the hall. The water temperature is around 5°C. The temperature of the air in my office is a little warmer than I want it to be, perhaps 25°C. I am interested in what happens to the temperature of the water as it sits on my desk.

There is little mystery. The temperature of the water will rise and approach 25°C. My cool container of water will eventually reach the same temperature as the room. I will denote by $T(t)$ the

Figure 2.1. My container of water. We will spend much of this chapter thinking about how the temperature of the water in this container changes.

temperature in degrees Celsius t minutes after I returned from the water cooler and put the container on my desk. Thus, the statement $T(10) = 13$ means that 10 minutes after I put the container on the desk the water is 13°C.

This situation is a dynamical system: we have something—the water temperature T—that's changing in time. We'd like to know the function $T(t)$. Determining an algebraic formula for $T(t)$ isn't easy. And, as I'll argue such a formula might not be necessary to understand the long-term dynamics of the equation. But it *is* easy to make a rough sketch of what $T(t)$ should look like. I've done so in Fig. 2.2. The water warms up: quickly at first, and then less and less quickly as the temperature of the water approaches room temperature.

Now let's use some physics and math to describe this situation. The physical law that describes how objects warm up or cool down is known as *Newton's law of cooling*. For our container of water in a 25°C room, the law takes the following form:

$$\frac{dT(t)}{dt} = -0.1(T(t) - 25). \tag{2.1}$$

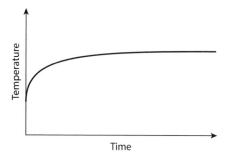

Figure 2.2. A sketch of the temperature of a cool container of water after it is placed in a warm room. Initially the water warms up quickly. The rate of warming decreases as the water gradually reaches room temperature. I've not indicated numerical values on this sketch, but we know that the temperature starts at 5°C and approaches 25°C, the temperature of the room. For the horizontal axis, perhaps the entire graph range would be one hour. We know from experience that after about an hour a cool container of water would be at about room temperature.

This equation can be used to determine $T(t)$, the temperature as function of time.[1]

But before we try to figure out $T(t)$, let's dissect Eq. (2.1). The term on the left, $\frac{dT(t)}{dt}$ is the derivative of the function $T(t)$. If you've studied calculus, you likely know (or knew) lots of tricks for evaluating derivatives. But for our purposes we won't need these tricks. It is sufficient to know that the derivative is the instantaneous rate of change of a function.[2] Here, the function's derivative

1. Newton's law of cooling is a standard topic in many thermodynamics books and is used as an example in many differential equations texts. Barnes and Fulford (2002, Chapter 9) is a particularly clear discussion of heat transfer and Newton's law. Understanding the physics of the cooling-water situation is not important for what is to follow. Newton's law of cooling applies to objects warming up— as is the case with the water bottle in this example—as well as objects that are cooling off.

2. The derivative is the central object studied in the first part of a sequence of courses on Calculus. For this book the most important thing is to have an understanding of what the derivative is—what it means and what it tells us. For

is the rate at which the water is warming up. For example, the statement $\frac{dT(10)}{dt} = 0.4$ means that after the water has been in the room for 10 minutes, its temperature is increasing at a rate of 0.4°C per minute.

We now turn our attention to the right-hand side of Eq. (2.1). The quantity $(T(t) - 25)$ is the difference between the temperature of the water and the temperature of the room, 25°C. The number 0.1 is a constant that is a property of the water container and the nature of its thermal contact with the air. Different situations, for example, a beer in a pint glass or coffee in Fa mug, will have a different constant than 0.1. In some cases it may be possible to determine the constant based on known material properties of the container. More commonly, it would be estimated from data. In this instance I experimented a bit and chose the value 0.1 because it seemed to give reasonable results. My goal here is to come up with a physically-motivated equation that we can study as a dynamical system, not to do careful thermodynamics or engineering.

To complete our dissection of the right-hand side of Eq. (2.1) we need to think about the minus sign. This term is here so that the object's temperature increases if it is colder than room temperature and decreases if it is warmer than room temperature. We'll see how this works out in a moment.

Putting it all together, Eq. (2.1) says that the rate of change of the temperature of the water is equal to -0.1 times the difference between the water's temperature and the room temperature, 25°C. For example, if the water's temperature is 10°C, then we can figure out the rate of change $\frac{dT}{dt}$ of the water by plugging in to the right-hand side of Eq. (2.1):

$$\frac{dT}{dt} = -0.1(10°C - 25°C) = 1.5. \qquad (2.2)$$

a conceptual discussion of the derivative, see e.g., Thompson and Gardner (1998), Feldman (2012, Chapter 28), or Feldman (2014, Unit 2.2).

Thus, if the temperature of the water is 10°C then its temperature is increasing at 1.5°C/minute. Note that the minus sign in front of 0.1 was needed to make the derivative positive, indicating an increasing temperature.

Equation (2.1) is a *differential equation*, an equation that states a relationship between a function and its derivative. Differential equations express a relationship between a quantity and its instantaneous rate of change, as we did when thinking about my warming container of cool water. The majority of the laws of physics are expressed as differential equations, and differential equations are used in a variety of ways throughout the sciences. I'll have more to say about the different ways that models— differential equations and others—are used in Sections 3.4 and 3.5 in the next chapter.

Our concern now, however, is a mathematical puzzle. Given a differential equation like Eq. (2.1), how do we go about finding the unknown function? There are several complementary ways to approach this. The first approach that I'll discuss is the one that students most often learn first.[3] I'll argue, however, that this approach is often unnecessary and may even be misleading, especially if one is considering differential equations from a dynamical systems point of view.

2.2 Exact Solutions

It is sometimes possible to obtain an exact solution to a differential equation. By exact solution, I mean an algebraic formula for the function that makes the differential equation true. Solving a differential equation entails finding an unknown *function*. In contrast, solving an "ordinary" equation, such as $2x + 4 = 10$ means finding the *number x* that makes the equation true. For Eq. (2.1),

---

3. It is not necessary to have studied differential equations before. Throughout this book I assume that readers have not had previous exposure to differential equations.

the differential equation describing my cooling water container, the unknown function $T(t)$ happens to be:

$$T(t) = 25 - 20e^{-0.1t}. \tag{2.3}$$

More generally, if the initial temperature is T_0 and not necessarily 5°C, then the solution is:

$$T(t) = 25 - (25 - T_0)e^{-0.1t}. \tag{2.4}$$

Verifying that the function given in Eq. (2.3) really does solve the differential equation, Eq. (2.1), is a fairly straightforward exercise in differential calculus. One simply plugs in $T(t)$ on both sides of the equation, evaluates the derivative, and simplifies to show that the equation is true. Similarly, one can show that $x = 3$ is a solution to the equation $2x + 4 = 10$ by plugging in $x = 3$, and using arithmetic to show that the equation is true.

Once one has an exact solution, one can do all sorts of things with it. For example, if I plug $t = 10$ into Eq. (2.3) I find that $T(10) \approx 17.6$, meaning that after 10 minutes the temperature of the water is around 17.6°C. One can also make a plot of the solution, but I'm going to hold off doing so for now. I want to show first in Section 2.4 how one can get a general sense of the shape of the solutions to a differential equation graphically, without using calculus or finding formulas.

Checking to see if a function solves a differential equation is fairly straightforward, but finding this function can be very difficult and often impossible. I'm not going to talk about *how* one would go about finding an exact solution, but in the next section I'll talk about why I don't want to talk about it.

2.3 ~~Calculus Puzzles~~

One way of approaching differential equations is to view them as a kind of calculus puzzle wherein one needs to find some unknown function whose derivative is some function of itself.

There are a variety of techniques and tricks one can employ to find the unknown function. If you've taken calculus, you probably recall that differentiating functions was not too difficult, but that anti-differentiating functions (also known as doing integrals) is sometimes much, much harder. It is not always clear which anti-differentiation technique one needs to use for a particular problem. To make matters worse, there are many functions whose anti-derivatives do not possess a closed-form solution. In such an instance an anti-derivative *exists*, but there is no formula for the anti-derivative that has a finite number of terms.

The situation is similar for differential equations, except perhaps a bit worse. There are many different classes or families of differential equations, and there are correspondingly many types of techniques and tricks for solving them. Moreover, the majority of differential equations do not possess closed-form solutions; there is no finite algebraic expression for them. Almost all non-linear differential equations are "unsolvable" in this sense.

Most differential equations textbooks and courses focus on analytic techniques—calculus tricks for solving differential equations. I think there are several drawbacks to this approach to differential equations. First, as previously noted, only a minority of differential equations are even amenable to such techniques. So a focus on analytically-solvable differential equations inevitably narrows one's scope. Studying techniques applicable to only a small subset of differential equations can be misleading. Non-linear equations are rarely solvable analytically, and as we shall see, non-linear equations can display chaotic behavior—a phenomenon not seen in linear equations. So viewing differential equations as calculus puzzles means that chaos is out of the picture. Moreover, I think that the calculus-puzzle view of a differential equation might obscure its nature as a dynamical system: a rule that tells how something changes in time.

I feel a bit remorseful for saying bad things about using calculus tricks to solve differential equations and titling this section

~~Calculus Puzzles~~ instead of Calculus Puzzles. So I'll end this section on a conciliatory note. Analytic solutions obtained via calculus tricks are of tremendous value, when they can be found. An *analytic* solution is one obtained with pencil-and-paper algebra and calculus, not via a numerical or approximate approach. It is often preferable to have a nice formula for a solution. And it turns out that for many of the central equations of physics, including the Maxwell and Schrödiner equations, it is usually possible to find analytic solutions, although sometimes those solutions are only approximate. Analytic solutions are also important because they can be used to check numerical solutions, as discussed in Section 2.5. Analytic methods can also be used to find approximate solutions or determine the asymptotic form of exact solutions. Finding analytic solutions is a lot of fun for certain types of people (including me). But for a book whose goal is to give an overview of dynamical systems, I don't think analytic solutions are worth the time or energy. In the last section of this chapter I'll give a few references for those who want to dig further into the analytic aspects of differential equations.

2.4 Qualitative Solutions

One can often learn a great deal about the general nature of solutions to a differential equation via some graphical techniques that I will illustrate in this section. We will use these techniques extensively in Chapter 6 when we look at bifurcations. To get us started, let's return to Eq. (2.1), describing the temperature T of a cool container of water as it warms up in a 25°C room. I'll re-write the equation here:

$$\frac{dT(t)}{dt} = -0.1(T(t) - 25). \qquad (2.5)$$

Let's make a plot of the right-hand side of this equation. The graph is a line, as shown in Fig. 2.3. Note that the horizontal axis is Temperature T, not time. The graph tells us how fast the water

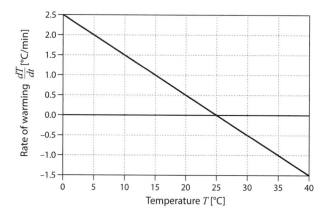

Figure 2.3. A plot of the right-hand side of Eq. (2.5).

is warming when the water is at a particular temperature T. For example, we see that if the temperature is 10°C, then the water is warming up at a rate of 1.5°C/min, which is what we found back in Eq. (2.2). If, say, the temperature was 35°C, as might be the case for a warm cup of coffee, then the rate of warming is −1°C/min. In this case, the temperature is decreasing; the coffee is cooling.

We can use the plot in Fig. 2.3 to determine the qualitative behavior of the temperature of the water. The initial temperature is $T = 5$°C. At this moment in time the water is warming up at a rate of 2°C/min. A little bit later, the water is warmer. The rate of warming is not as large as before, but it is still positive, so the temperature continues to increase. The story continues. The temperature continues to increase while the rate of warming decreases. The temperature gets closer and closer to 25°C. This was the behavior I sketched by hand in Fig. 2.2.

Let's take a more global view: what is the long-term behavior of *all* solutions to the differential equation, Eq. (2.5)? All temperatures less than 25°C will increase and approach 25°C. We know this from common-sense physics—cool objects eventually

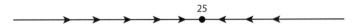

Figure 2.4. The phase line for the differential equation Eq. (2.5).

reach room temperature. But we can also see this in Fig. 2.3. If
the temperature T is less than 25 Celsius, then the rate of warming is positive; the temperature increases. And if the temperature
is greater than 25 Celsius, then the rate of warming is negative,
and the temperature decreases. This behavior is summarized in
Fig. 2.4, which is referred to as the *phase line* for the differential
equation.

A phase line is a useful graphical way to summarize the behavior
of differential equations of the form of Eq. (2.5). It shows the
long-term behavior of solutions to the differential equations for
all possible initial conditions. In Fig. 2.4 we see that any initial
temperature less than 25 increases and approaches 25. And any
initial temperature greater than 25 decreases and approaches 25.
The phase line is a one-way street. At any point on the phase line
all the solutions move in the same direction.

The phase line shows us that there is a stable, or attracting, fixed
point at $T = 25$. All temperatures are pulled toward 25. Fixed
points of differential equations are more commonly referred to
as *equilibria*. For a differential equation an equilibrium (or fixed
point) is one where the derivative is zero, since a zero derivative indicates that a quantity is unchanging. The equilibrium
condition is:

$$\frac{dT}{dt} = 0. \tag{2.6}$$

For Eq. (2.5), the derivative is zero when

$$-0.1(T - 25) = 0. \tag{2.7}$$

As expected, we see that the solution to this equation occurs when
$T = 25$; if I plug in $T = 25$, the Eq. (2.7) is true. Thus, $T = 25$ is
a fixed point.

Phase lines such as Fig. 2.4 are a clear way to indicate the global behavior of a differential equation—a phase line shows the fixed points and their stability. From a phase line, one can immediately see the long-term behavior of all initial conditions. However, a phase line does not give us a full specification of $T(t)$, the temperature as a function of time t. One consequence of this is that it is not possible to use a phase line to tell how fast a trajectory moves. For our water example, this means that one cannot tell from Fig. 2.4 how quickly the temperature approaches 25°C. If we desire this sort of information, we will need another approach to solving the differential equation.

2.5 Numerical Solutions

In this section I'll discuss a way to determine, to arbitrary accuracy, the solution $T(t)$ to a differential equation like Eq. (2.5). The method I'll introduce is known as *Euler's method*. I like this method because unlike the calculus puzzle approach, it gets right at the heart of what a differential equation is: a rule that tells one the instantaneous rate of change of a function. I'll start by illustrating Euler's method on the familiar warming-container-of-water example.

We know that $T(0)$, the temperature of the water immediately after it is brought into the room, is 5°C. What happens next? We don't know any subsequent values for the temperature, but we do know how fast the water is cooling off at the initial moment $t = 0$. The rate of change of the temperature is given by the differential equation:

$$\frac{dT(t)}{dt} = -0.1(T(t) - 25). \qquad (2.8)$$

Since initially the temperature is 5, we plug 5 into the right-hand side of Eq. (2.8) and obtain

$$\frac{dT(0)}{dt} = -0.1(5 - 25) = 2. \qquad (2.9)$$

So the moment the water is brought into the room it is warming up at a rate of 2°C/min.

Suppose we want to know $T(2)$, the temperature of the water 2 minutes after it has been in the room. We use the rate of warming at $t = 0$ to obtain:

$$T(2) = 5°C + (2°C/min) \times 2\,min$$
$$= 5°C + 4°C = 9°C. \qquad (2.10)$$

I've included units on the terms in this equation to make it clearer what is going on. The temperature increases at a rate of 2°C/min for 2 minutes, so the temperature increases 4°C during these two minutes. Since the water started at 5°C, after two minutes it is at 9°C.

I hope this this is all fairly straightforward. However, there is a problem with Eq. (2.10): it is based on a lie. The rate of warming is *not* constant; it continually decreases as the water warms up. Since the rate of change is not constant at 2°C/min during the two-minute interval, we cannot say that the temperature gain was 2°C/min $\times 2$ min $= 4°C$. What do to? We *pretend* that the temperature is constant over these two minutes.

Carrying on, then, we assert that $T(2) = 9°C$ (even though it doesn't really), and then can ask about the temperature when $t = 4$. At $t = 2$ when the temperature is 9°C, how fast is the water warming up? We ask the differential equation. Plugging $T = 9$ in to Eq. (2.8), we find:

$$\frac{dT(2)}{dt} = -0.1(9 - 25) = 1.6. \qquad (2.11)$$

We can use this to figure out the temperature at $t = 4$. Proceeding as we did before:

$$T(4) = 9°C + (1.6°C/min) \times 2\,min = 12.2°C. \qquad (2.12)$$

Again, this equation is based on a lie. In fact, it's based on two lies. One lie is that the rate of warming is constant over the two-minute interval from $t = 2$ to $t = 4$. This isn't true since, as previously noted, the rate of warming changes as the water warms up. The second lie is that $T(2) = 9$. This isn't true, since Eq. (2.10) was based on the lie that the rate of warming is constant from $t = 0$ to $t = 2$.

It is reasonable to get worried. Lies are accumulating. But let's keep our head down and move forward one more step. What is $T(6)$, the temperature six minutes in? We need to know how fast the water is warming up at $t = 4$ when it is $12.2°$C. To get the rate, we ask the differential equation. Plugging $T = 12.2$ in to Eq. (2.8), I find $1.28°$C/min. This then yields a value of $T(6) = 14.76°$C. We can continue in this fashion, pretending the rate of warming is constant for two-minute intervals, and calculating $T(8)$, $T(10)$, and so on. The results of this method are shown in the second column of Table 2.1. (You might want to grab a calculator and check my results for $T(8)$ and $T(10)$ to make sure you are following the logic.)

The method I have just led us through is known as *Euler's method*. To implement the method one needs to choose a step size Δt, the interval of time over which we pretend that the growth rate is constant. In this example, I chose $\Delta t = 2$. Once these choices are made, roll up your sleeves, grab a calculator, and away you go. Or, program a computer to do the work for you.

Let's step back and assess our Euler's method solution. Is it any good? For this differential equation we happen to have a formula for the exact solution, Eq.(2.3). The exact results are shown in the right-most column of Table 2.1. The Euler results do not agree with the exact solution. This is scarcely surprising—after all, the Euler "solution" is based on a series of lies. Note that the Euler solutions are consistently too large; they give the temperature as being warmer than it actually is. It makes sense that Euler's method gives us over-estimates in this case. When we used Euler's method we pretended that the rate of warming was constant during each two-minute interval. In reality, the rate of warming continually

Time	Euler $\Delta t = 2$	Euler $\Delta t = 1$	Exact
0	5	5	5
1		7	6.90
2	9	8.80	8.63
3		10.42	10.18
4	12.2	11.88	11.59
5		13.19	12.87
6	14.76	14.37	14.02
7		15.43	15.07
8	16.81	16.39	16.01
9		17.25	16.87
10	18.45	18.02	17.64

Table 2.1 Solutions to the differential equation Eq. (2.8).

decreases as the water gets closer to room temperature, a reality ignored by Euler's method. Accordingly, in this example our approximates are always larger than the exact value.

As noted, Euler's method yielded an approximate answer because it pretends that the warming rate is constant over a two-minute interval. We can make the error smaller by using a shorter interval. Let's pretend that the warming rate is constant over one minute instead of two minutes. This is still a lie, but it is less of a lie than before. The rate of warming changes less over a one-minute interval than it does over a two-minute interval. Euler's method proceeds as before, except we use $\Delta t = 1$. Let's do the first few steps. Initially, we know that $T(0) = 5$. We determine the rate of warming by asking the differential equation, Eq. (2.8). Plugging in $T = 5$ to the right-hand side of the differential equation, we are told that when $T = 5$, the water is warming up at a rate of 2°C/min. So, $T(1)$, the temperature after one minute is:

$$T(1) = 5°C + (2°C/\text{min}) \times 1\,\text{min} = 7°C. \qquad (2.13)$$

This result is shown in the third column of Table 2.1.

Let's do one more step. We know that $T(1) = 7°C$, and we want to know $T(2)$, the temperature after two minutes. What is the rate of warming when $T = 7°C$? We ask the differential equation. Plugging $T = 7$ in to the right-hand side of Eq. (2.8), we find:

$$\frac{dT(1)}{dt} = -0.1(7 - 25) = 1.8. \qquad (2.14)$$

We then use this to determine $T(2)$:

$$T(2) = 7°C + (1.8°C/\text{min}) \times 1\,\text{min} = 8.8°C. \qquad (2.15)$$

One can keep going, pretending the rate of warming $\frac{dT}{dt}$ is constant for one-minute intervals, to obtain the results collected in Table 2.1. Note that the Euler approximations with $\Delta t = 1$ are again overestimates, but they are closer to the exact values, as one would expect. Pretending the rate of cooling to be constant over a one-minute interval is less of a lie than pretending it is constant over a two-minute interval.

In Fig. 2.5 I have plotted the exact solution to Eq. (2.8) along with the Euler approximations for $\Delta t = 2$ and $\Delta t = 1$. One can see that the Euler method gives a fairly good approximation to the exact solution, despite the fact that the method is based on a lie. One can make the results for Euler's method more and more accurate—closer to the exact solution—by letting Δt get smaller and smaller. This can get tedious to do by hand, but computers can do this with ease. For example, if I set $\Delta t = 0.01$, my ordinary desktop computer can apply Euler's method and plot the result in under one second. The resulting curve is indistinguishable from the exact solution.

Euler's method yields what is referred to as a *numerical solution*. This means that the solution that results is a list of numbers and

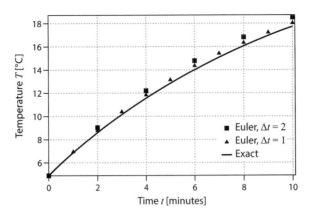

Figure 2.5. A plot of the exact solution for the differential equation, Eq. (2.8), describing the temperature of a warming container of water. Also shown as squares and triangles, respectively, are the approximate solutions obtained via Euler's method for $\Delta t = 2$, and $\Delta t = 1$.

not a formula. Numerical solutions like Euler's method are not exact. They are based on lies—the constancy of the rate of change over a time interval Δt. This lie can be made smaller and smaller by letting Δt get smaller and smaller. As one does so, the Euler solution gets closer and closer to the exact solutions. Mathematically, one would say that the Euler solution *converges* to the exact solution.

I'll have a little more to say about numerical solutions to differential equations in Section 2.7. But before doing so, I want to collect the results of the last several sections and look at some complementary ways of visualizing and thinking about solutions to Newton's law of cooling, Eq. (2.8).

2.6 Putting It All Together

In this chapter we have been studying the differential equation that describes how the temperature of a container of water changes when it is placed in a 25°C room. Writing that equation one more time:

$$\frac{dT(t)}{dt} = -0.1(T(t) - 25). \qquad (2.16)$$

A differential equation is a dynamical system: a rule that tells us how something changes in time. Given the starting value (in our example 5°C) and the rule, we can figure out $T(t)$, the temperature as a function of time. In the preceding several sections I presented a number of ways do this.

If one plots the right-hand side of the differential equation, as I've done in the top plot in Fig. 2.6, one can see immediately the regions in which the function $T(t)$ is increasing or decreasing. One can also see the fixed points; they occur when the derivative is zero. From this information, one gets a sense of the dynamics of the differential equation: the fixed points and their stability. One can then infer the long-term behavior of any initial condition. This information is usefully summarized with a phase line. In the top of Fig. 2.6 I've drawn the phase line directly on the graph of the right-hand side of Eq. (2.16). The phase line reveals the long-term dynamical behavior of the differential equation.

In many instances it is useful to plot solutions as a function of time. I've done this in the bottom plot of Fig. 2.6. I've chosen three starting temperatures, 5, 15, and 35°C and shown the temperature as a function of time for each. Not surprisingly, in all cases the temperature of the water approaches room temperature, 25°C. One could obtain plots for $T(t)$ either via Euler's method or an exact algebraic solution, if one exists.

At this point I'm finally ready to stop discussing the temperature of the container of water. I'm aware that I may have crossed the line into repetition, but my hope is that carefully treating an almost-too-simple example has helped to make it clear what differential equations are and how to think about them.[4] The

4. If you'd like to see another example, you can skip ahead to Section 6.1, in which I analyze a different differential equation.

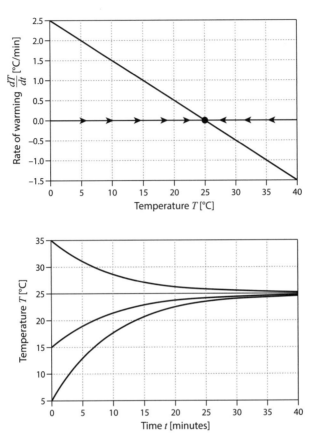

Figure 2.6. The top plot shows the right-hand side of Eq. (2.16). The phase line for the differential equation is drawn on the horizontal axis. Whenever $\frac{dT}{dt}$ is positive (negative) the temperature increases (decreases). There is a stable fixed point at $T = 25°C$. The bottom plot shows solutions to Eq. (2.16), obtained via Euler's method, for three different initial temperatures: 5, 15, and 35°C. The stable equilibrium value of $T = 25$ is indicated by the thin horizontal line.

rest of this chapter contains some more general comments and observations about differential equations.

2.7 More about Numerical Solutions

There are two other issues around numerical solutions that I'd like to discuss. First, you should know that Euler's method is not the most efficient algorithm for solving a differential equation. Efficiency in this context refers to the amount of time or memory a computer needs in order to obtain a solution to a given accuracy. A family of methods known as *Runge–Kutta* (RK) methods are more efficient than Euler and are implemented as part of the differential equation solvers that are part of matlab, scipy (the scientific computing package for python), and many other languages and platforms. RK methods are similar to Euler's method, but rather than use the derivative at a single point to determine the pretend-constant rate of change over a time interval Δt, a weighted average of the derivatives at several different times is used.[5]

You might ask, then, why I have focused on the less efficient Euler method rather than the more efficient RK method. I presented Euler's method because it is conceptually clean. It gets right to the heart of what a differential equation is: a rule for how something changes in time. The rule is indirect, in that it gives us the rate of change, rather than the thing itself. Further complicating matters, the rate of change (i.e., the derivative) usually is continually changing. Euler's method lets us go from this continually changing derivative to the function.

The second reason I introduced Euler's method and not RK is that if you know how Euler's method works, then I think you

---

5. Additionally, most of these built-in differential-equation solvers use an adaptive step size algorithm. This is an algorithm that changes the step size Δt depending on the characteristics of the equation. The step size needs to be smaller in intervals where the derivative is changing rapidly, since here the lie that the derivative is constant over an interval is farther from the truth than for intervals where the derivative changes slowly.

have a pretty good conceptual picture of how RK works. Often in this book—and in other books and in the scientific literature—you will read simply that a solution to a differential equation was obtained numerically. You can picture a computer doing a slightly more clever version of Euler's method and you'll have the right idea what is going on inside the "black box" of the RK algorithm.

The second issue I want to discuss about numerical solutions is a bit more philosophical. Some might object to calling the results of Euler's method a *solution* to a differential equation. They might argue that the only way to truly solve a differential equation is to find an algebraic formula that exactly solves it. I disagree. I think it is perfectly legitimate to say that Euler or RK methods can solve a differential equation. Yes, numerical methods give numbers, but a list of numbers is a perfectly fine way to specify a function. And yes, the numbers are only an approximation to the exact solution, but in almost all cases we can make this approximation as accurate as we want by decreasing the step size.

RK methods are not terribly difficult, but they're a bit involved and so beyond the scope of this short book. A description of them can be found in almost any numerical methods text. The discussion in Section 8.1 of Newman (2012) is particularly clear. See also Chapter 16 of Press et al. (1995) for a detailed and highly readable account of a variety of algorithms for solving differential equations along with cautionary remarks about some the perils of solving differential equations numerically.

2.8 Notes on Terminology and Notation

In this section I'll discuss some different categories of differential equations and introduce some terminology. First, a few words about alternate notation for derivatives. I have used the notation $\frac{dx}{dt}$ to denote the derivative of the function x with respect to time. This form for writing the derivative is known as Leibniz's notation.

Another very common notation, known as Lagrange's notation, is x'. The two notations are equivalent and are used interchangeably:

$$x' = \frac{dx}{dt}. \tag{2.17}$$

Another common notation, due to Newton, is to place a dot over a variable to indicate a derivative with respect to time. That is,

$$\dot{x} = \frac{dx}{dt}. \tag{2.18}$$

This notation is most common in physics and engineering. I will not use dot notation in this book, but it is likely you will encounter it elsewhere. Note that Newton's dot notation is only used to refer to time derivatives.

So far our discussion of differential equations has focused on one example, Newton's law of cooling, Eq. (2.16). This differential equation is of the form:

$$\frac{dx}{dt} = f(x). \tag{2.19}$$

Here x is an unknown function of time: $x(t)$ is what we're solving for. The right-hand side is some function involving x, the unknown function. (In the warming water example the unknown function was $T(t)$, not $x(t)$.) Equations of the form of Eq. (2.19) are known as *autonomous* differential equations. The idea is that the system does its own thing independent of time—and hence is autonomous. The rate of change depends only on the value of x, not on the time. In the warming water example, the rate at which the water warmed depended only on the water temperature T, and not on the time t.

Another type of differential equation has a time-dependence on the right-hand side:

$$\frac{dx}{dt} = f(x, t). \tag{2.20}$$

Equations of the form of Eq. (2.20) are called *non-autonomous* differential equations. The following equation is of the form of Eq. (2.19), and thus is autonomous:

$$\frac{dx}{dt} = 2x(1-x).\tag{2.21}$$

This equation:

$$\frac{dx}{dt} = 2x(1-x) + \sin\left(\frac{3t}{\pi}\right),\tag{2.22}$$

on the other hand, is of the form of Eq. (2.20) and thus is non-autonomous, because it has a *t* on the right-hand side. In this book I will cover only autonomous differential equations.

The *order* of a differential equation refers to the highest order of the derivative that appears in it. Equations (2.21) and (2.22) are both *first-order* equations, because the highest derivative that appears is the first derivative. The equation

$$\frac{d^3x}{dt^3} = x + x^2 + 7x\frac{dx}{dt},\tag{2.23}$$

is a third-order equation. A differential equation is called *ordinary* if it contains only ordinary, and not partial, derivatives. Ordinary differential equations often go by the poetic abbreviation ODE. Partial differential equations are known by the less poetic PDE.

I'll end this section by introducing one more bit of terminology. In order to solve a differential equation of the form of Eq. (2.20), one needs to specify an initial condition. Thinking of Euler's method for the warming water example, in order to figure out the solution, we need to know not just the differential equation but also the starting temperature of the water. A differential equation along with an initial condition is collectively referred to as an *Initial Value Problem* and is often abbreviated IVP.

2.9 Existence and Uniqueness of Solutions

In mathematics it is nice to know that a solution exists before one sets off trying to find it. Do all differential equations have solutions? Might they have more than one solution? Let's first think about these questions in the context of iterated functions, introduced in the previous chapter. Iterated functions are dynamical systems of the form:

$$x_{n+1} = f(x_n) . \tag{2.24}$$

It is a simple matter to determine the itinerary for an initial condition. One starts with the initial condition and applies the function over and over again. In the context of differential equations, one would say that we've found a solution to the equation—we've solved for the unknown itinerary, just as when faced with a differential equation we seek to solve for the unknown function.[6]

In any event, the "solution" to Eq. (2.24) exists. How could it not? As long as the function $f(x_n)$ on the right-hand side doesn't behave pathologically, perhaps by requiring a division by zero, there is no way that you'll get stuck as you carry out the process of determining the itinerary. Moreover, this itinerary is unique. There is one and only one itinerary for each function f and each initial condition x_0. If you and I each start with the same initial condition and enter numbers correctly on a reliable calculator, we are guaranteed to compute the exact same itinerary.

The situation is the same for autonomous first-order differential equations of the form of Eq. (2.19). If the function $f(x)$ is continuous and differentiable in a region that includes the initial condition $x(0)$, then there exists a unique solution. The existence and uniqueness of solutions to differential equations is a standard

---

6. Although in practice one almost never says that one solves Eq. (2.24), because "solving" seems like too grand a word to describe just plugging a number in to your calculator and applying a function to it over and over.

part of most differential equation texts. For good discussions of some of the issues around existence and uniqueness, see Section 1.5 of Blanchard, Devaney and Hall (2011) or Section 7.2 of Hirsch, Smale and Devaney (2004). For a proof of existence and uniqueness of solutions, see for example, Section 3.3 of Robinson (2012).

The existence and uniqueness theorems basically say that as long as the right-hand side of the differential equation is well behaved, there will exist a unique solution. I like to think of existence and uniqueness in the context of Euler's method. If one applies Euler's method with smaller and smaller time steps Δt, one will be led closer and closer to the exact solution. How could one not be? As long as the derivative changes continuously and not with any sudden jumps, there is only one possible solution.

The upshot of all this is that differential equations of the form of Eq. (2.19) are well-posed problems. They have one and only one solution. I hope this result is fairly intuitive. Viewed as a dynamical system, a differential equation is an unambiguous rule that specifies how a quantity changes. Given an initial condition and this unambiguous rule, there is one and only one trajectory possible. The initial condition and the equation uniquely determine the solution.

2.10 Determinism and Differential Equations

Before this chapter draws to a close, I want to present a result that places strong limitations on the types of behavior possible for autonomous differential equations of the form

$$\frac{dx}{dt} = f(x), \qquad (2.25)$$

the type of equation we have been examining in this chapter. A solution to this differential equation can not oscillate in any way. This is perhaps most easily seen via a counterexample. In Fig. 2.7 I have shown a function $x(t)$. Could it be a solution to Eq. (2.25)?

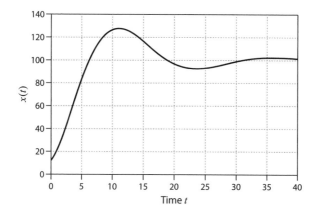

Figure 2.7. A plot of a function $x(t)$ that cannot possibly be a solution to an autonomous differential equation of the form $\frac{dx}{dt} = f(x)$.

The answer is "no". This can be seen as follows. Consider, say, $x = 120$. The function $x(t)$ has this value at two instants, $t \approx 7$ and $t \approx 14$. At $t \approx 7$, $x(t)$ is increasing, so its derivative is positive. And at $t \approx 14$, $x(t)$ is decreasing and hence has a negative derivative.

So what we have just seen is that when $x = 120$, the derivative of x, dx/dt, could be either positive or negative. But Eq. (2.25) says that dx/dt is a function *only* of x and not of t. Thus, the same value of x should always yield the same value of the derivative dx/dt. We are thus forced to conclude that the function $x(t)$ shown in Fig. 2.7 cannot possibly be a solution to an autonomous differential equation of the form of Eq. (2.25).

The same story will hold for any other $x(t)$ that oscillates (i.e., that both increases and decreases). We thus conclude that autonomous differential equations of one variable cannot oscillate. The type of behaviors seen in solutions to autonomous differential equations is thus rather limited. A solution can either increase or decrease, but not both. It can approach an equilibrium, or it can tend toward positive or negative infinity. That's it. This result, thought of in terms of phase lines, simply says that the phase line is a one-way street. At every point on the phase line (except for fixed points) solutions either increase or decrease, but not both. We shall see in a few chapters, however, that systems of differential equations can have much more interesting and complex behavior.

2.11 Iterated Functions vs. Differential Equations

In this chapter and the last we looked at two types of dynamical systems: iterated functions and differential equations. These are different classes of models. For iterated functions, time is discrete; the result of iterating a function is a time series that consists of a list of values. For example, applying the squaring function to the seed $x_0 = 2$ yields the orbit $2, 4, 16, \ldots$. The system has the value of 2 at time 0, 4 at time 1, 16 at time 2, and so on. It does not have a value at intermediate times. It is meaningless to ask about the value at time 0.5. A consequence of this is that values of the orbit jump from one to the next. For example, if x_0 is 2 and x_1 is 4, that does not mean that x passes through the values between 2 and 4.

In contrast, the solution $x(t)$ to a differential equation is defined for all t, not just integer values of t. Additionally, the solution to a differential equation will generally be continuous. For example, if a differential equation's solution $x(t)$ has a value of 2 and a little while later is at 4, it must have visited all values between 2 and 4 along the way.

The upshot is that differential equations and iterated functions are different kinds of dynamical systems with qualitatively different properties. For example, as we shall see, iterated functions can exhibit chaos, but one-dimensional differential equations can not. Chaos for differential equations is only possible in three or more dimensions.

That said, there are some connections between differential equations and iterated functions. First, as you have perhaps noticed, in the process of solving a differential equation via Euler's method, we have reduced a differential equation to an iterated function. Like the iterated functions of Chapter 1, Euler's method is a rule that uses the value of a function at one time step to determine the function's value at the next time step. Euler's method is an iterated function that can emulate any well-behaved differential equation. The converse is not true. Given an arbitrary iterated function, it is not always possible to emulate it with a differential equation.

Note that Euler's method is used to find approximate solutions to a continuous function. Thus, if Euler's method determined that the temperature of a cup of coffee was 35 degrees at $t = 1$ minute and 33 degrees at $t = 2$ minutes, then we can be sure that at some instant between minutes one and two the cup of coffee was 34 degrees. As noted, in general for iterated functions there is no expectation of continuity, nor is it meaningful to consider the value of the solution at non-integer time values.

A second connection between iterated functions and differential equations is that iterated functions are often used to analyze differential equations. We will see examples of this in Sections 9.4 and 9.6. Rather than working directly with the differential equation, it is sometimes easier to use the differential equation to construct an iterated function that captures some feature of the differential equation. Properties of the differential equation can then be deduced from the properties of this function.

Third, iterated functions exhibit most of the dynamical phenomena that differential equations do, yet are usually considerably easier to study. So studying iterated functions is a great way to learn about many of the key concepts and results from dynamical systems. I will take this approach here. In Chapters 4 and 5 we will see how chaos and the butterfly effect arise in iterated functions. Chaos in differential equations is covered later, in Chapter 9.

2.12 Further Reading

Differential equations are a standard part of the mathematics curriculum; it is one of the first math classes offered beyond calculus and is taken by all math, physics, and engineering majors, and many others. As such, there are a range of strong textbooks on differential equations with different styles and points of emphasis. One of my favorites is the text by Blanchard, Devaney, and Hall (2011). It is quite clear and readable and takes a dynamical systems point of view, striking a balance among qualitative techniques, numerical methods, and analytic methods. There are also a number of books on mathematical modeling and mathematical biology that contain excellent discussions of differential equations and their applications. These include: Britton (2005), Edelstein-Keshet (2005), Ellner and Guckenheimer (2006), Garfinkel et al. (2017), Mangel (2006), and Vandermeer and Goldberg (2013).

3

INTERLUDE: MATHEMATICAL MODELS AND THE NEWTONIAN WORLDVIEW

3.1 Why Isn't This the End of the Book?

At this point you may well be wondering why this isn't the end of the book. After all, I've introduced two types of dynamical systems: iterated functions and differential equations. For both I've discussed how to solve for the future behavior given the function or differential equation and an initial condition. And I hope to have convinced you that solving for the future behavior is not a difficult task, at least if one has access to a computer that can implement Euler's method for differential equations. Given a dynamical system we can determine the fixed points and their stabilities, and we can figure out the long-term behavior of any initial condition. These simple dynamical systems have simple behavior and it is pretty simple to figure out this behavior. End of story?

Actually, it turns out that simple dynamical systems such as those introduced in the previous two chapters most definitely do *not* always possess simple dynamical properties. Dynamical systems can exhibit surprisingly complex behavior, exhibit sudden transitions, and can also evolve in a way that is indistinguishable from random. This was first glimpsed by Henri Poincaré in the late 1800s, but it wasn't until the 1970s and 80s that the full richness of dynamical systems was appreciated and understood. The remainder of this book is devoted to giving an overview of the key lessons and realizations to emerge from the study of dynamical systems. This will begin in the next chapter, where we'll encounter chaos: unpredictable behavior produced by a deterministic system. In this chapter I'll touch upon a number of philosophical and conceptual issues that will help to set the stage.[1]

3.2 Newton's Mechanistic World

In 1687 Isaac Newton published *Philosophiae Naturalis Principia Mathematia* (Mathematical Principles of Natural Philosophy), usually referred to simply as the *Principia*. This work is often seen as the culmination or apotheosis of the scientific revolution. In the *Principia*, Newton gave a unified account of gravity, showing that the laws of gravity that hold on earth apply equally well to the moon, sun, and planets. In addition to co-inventing calculus, Newton put forth three universal laws of motion describing how objects move: not just here on earth but everywhere in the universe. Changes in motion are a consequence of forces acting on an object. Given knowledge of these forces, one can calculate the future position of the object, just as, given Newton's law of cooling, Eq. (2.1) in the previous chapter, one can calculate the future temperature of the water.

The *Principia* may seem like ancient history, but I bring it up because the ideas Newton put forth form the basis for a view of the world that I think is in many ways still with us. In the Newtonian picture the world is mechanistic and material. The universe is made up of stuff that exerts forces on other stuff. Objects move not because of spirits or desires, but because objects interact with other objects, and those interactions can be described by universal rules—universal in the sense that the same unchanging rules

1. Section 3.2 and 3.3 closely follow Feldman (2012, Chapter 8).

apply everywhere in the universe. The world is much like a giant machine. In the Newtonian framework the world is also mathematical. The laws of nature—the rules governing how material objects interact and behave—are expressed in mathematical form. The business of science, then, is to figure out what mathematical form these laws take.

Underneath all this is the assumption that the world is governed by rules or laws in the first place, and that these laws tend to have a simple mathematical form. Further, the presumption is that systems described by these simple laws have simple behavior. This presumption is, to my knowledge, almost always implicit or unspoken. It seems just to be just common sense that simple mathematical systems will behave simply. In this view, complexity arises in systems of very many interacting objects, or perhaps in situations where, for some reason, the mathematical laws or rules are not simple. (For further discussion of some of the assumptions of the Newtonian worldview see Diacu and Holmes (1999) and Stewart (2002).)

3.3 Laplacian Determinism and the Aspirations of Science

So the Newtonian world is machine-like. The world is vast and seems complicated, but ultimately made of simple gears and cogs. This leads to a certain optimism; the world should be understandable. We can't grasp it all at once, of course, but we can get ever closer to the true picture of things. But this mechanistic view of the world also leads to some puzzles. If the world really is a machine, perhaps the future has already been written. After all, the future is just a predictable and logical consequence of the present state of affairs.

In such a world, is there any choice or volition? I might wind up a toy robot, set it on a table, and let it go. The robot would stagger forward and fall off the table. It would be absurd to say

Figure 3.1. Are we all just complicated toy robots? (Image by D.J. Shin (Own work) [CC BY-SA 3.0 (http://creativecommons.org/licenses /by-sa/3.0) or GFDL (http://www.gnu.org/copyleft/fdl.html)], via Wikimedia Commons.)

that the robot *chose* to walk off the table. But in the Newtonian view, admittedly taken to the extreme, perhaps I am also a toy robot, albeit (hopefully) a much more complicated and complex one (see Fig. 3.1). If so, then perhaps I no more chose to write this book than the toy robot chose to fall off the table. Is there room for free will in a deterministic and mechanistic universe?

A related set of issues was laid out in 1814 by Pierre-Simon de Laplace in a now-famous passage:

> We may regard the present state of the universe as the effect of its past and the cause of its future. An intellect which at a certain moment would know all forces that set nature in motion, and all positions of all items of which nature is composed, if this intellect were also vast enough to submit these data to analysis, it would embrace in a single formula the movements of the greatest bodies of the universe and those

of the tiniest atom; for such an intellect nothing would be uncertain and the future just like the past would be present before its eyes. (Laplace, 1951)

This "vast intelligence" is now referred to as *Laplace's demon.*

There are a number of technical arguments showing "convincingly that neither a finite, nor an infinite but embedded-in-the-world intelligence can have the computing power necessary to predict the actual future, in any world remotely like ours (Hoefer, 2016).[2]" The supreme intelligence of Laplace's demon is clearly an unobtainable ideal. However, I think that one way of viewing science is that it is an ongoing effort to get closer and closer to Laplace's demon so that one can make accurate and reliable predictions about the future. As I have suggested elsewhere (Feldman, 2012, Section 8.3), there are three things scientists must do in order to approach Laplace's demon:

1. One needs to know the rules or laws of nature—how it is that the current state of the world determines the next state.

2. One needs to have accurate measurements of the current state of the world, since this is needed to predict the future state.

3. One needs sufficient computing power and memory to store an accurate representation of the current state of the world and carry out the calculations necessary, applying the laws of nature, to predict the future state.

This is surely an over-simplified view of science, but I do think it captures many aspects of how science functions—or at least how we think about science functioning. Scientific progress often occurs via advances along one or more of these three fronts: a better understanding of the rules of nature; more careful or accurate

2. These arguments basically amount to demonstrating that a computer needed to predict the future of the universe would need to be bigger than the universe.

measurements, perhaps made possible by some new measurement or imaging technology; and bringing greater mathematical ingenuity or computing power to bear on a problem. The study of dynamical systems has not, in my view, rendered obsolete this sketch of science and scientific progress, but it has complicated it considerably. For example, we shall see in subsequent chapters that simple deterministic systems can be fantastically difficult to predict, requiring insanely accurate measurements and vast computing power.

In this view of science much emphasis is placed on figuring out the rules—determining the laws of nature. The assumption is that once this is done correctly, the work of science is essentially done, and the phenomenon is understood. The study of dynamical systems challenges this assumption. We will see that dynamical systems, even very simple ones, possess behaviors that do not seem to immediately follow from the equations themselves.

Finally, I should note that there is wide range of practices that are considered science. Natural history, botany, genetics, chemistry, quantum physics, observational astronomy, and string theory all are generally viewed as science, although a reasonable case can made that some aspects of string theory fall outside of science. But nevertheless, there clearly is a diverse range of scientific practices and explanatory modes, not all of which emphasize prediction in the way that Laplace's demon does. In a similar vein, mathematical models find a range of uses in the sciences, a topic I discuss in the next few sections.

3.4 Styles of Mathematical Models

Just as there is a range of practices that are considered science, there is also a diversity of ways that mathematical models and equations are constructed and put to use. Many of us, I suspect, first used scientific equations to describe reality in physics or chemistry classes. In this setting equations are taken literally: the pressure of the ideal gas *is* 1.25 atmospheres, or the (x, y) coordinates of a projectile at a particular moment in time *really are* ($x = 25$ meters, $y = 38$ meters). In these applications, one rarely thinks one is doing modeling. The equations are viewed as faithful representations of real objects and processes. Often, the impetus for the use of equations in this context is to make predictions. We can predict how hot a gas will be if its pressure is increased by a certain amount, or we can figure out at what angle we should launch a projectile in order for it to land at a particular location.

Quite often in physics, and elsewhere in the sciences as well, understanding is equated with quantitative prediction. There are, however, other ways that equations are used to model processes and phenomena. Rather than a faithful or mechanistic accounting of the entirety of a phenomenon, one might instead wish to highlight just one or two essential aspects of the object or process being studied. Models like these are more like a caricature, a simple sketch designed to highlight, or perhaps even exaggerate, some feature of a phenomenon. The artist who draws a successful caricature of a subject surely has understood the subject and is able to convey this understanding to others through a sketch, even though the sketch probably is not helpful in predicting the future behavior of its subject.

One might think that a truer and more complete model of a phenomenon would be preferable, but that is not always the case. It depends on one's goals. For example, suppose one wants a guide to identify birds eating at your birdfeeder or that you see when hiking. One might expect that a field guide filled with photographs would be ideal for this purpose. After all, photographs are realistic, accurate representations.[3] However, good sketches or drawings are far superior for the purpose of identifying birds.

---

3. Of course, it's a bit more complicated than this, even before the current era in which manipulation of digital images is easy and routine. Composition, light, and resolution all can make a significant difference in the appearance of a photograph, and a skillful photographer can manipulate this to produce the desired effect.

Roger Tory Peterson is one of the preeminent naturalists, illustrators, and authors of the twentieth century. The following passage, written by his wife Virginia Marie Peterson in the introduction to one of his many field guides for birds, speaks to the value of illustrations:

> A drawing can do much more than a photograph to emphasize the field marks. A photograph is a record of a fleeting instant; a drawing is a composite of the artist's experience. The artist can edit out, show field marks to best advantage, and delete unnecessary clutter. He can choose position and stress basic color and pattern unmodified by transitory light and shade. A photograph is subject to the vagaries of color temperature, make of film, time of day, angle of view, skill of the photographer, and just plain luck. The artist has more options and far more control even though he may at times use photographs for reference. This is not a diatribe against photography; Dr. Peterson was an obsessive photographer as well as an artist and fully aware of the differences. Whereas a photograph can have a living immediacy a good drawing is really more instructive (Peterson, 1989, page xi).

Photographs and illustrations are both models of something: a bird. Actually, it's a bit more abstract than this— in a field guide they are models not of an individual bird but of an entire species, within which there may be considerable variation. An illustration in a field guide is thus a composite representation, capturing and highlighting what is common across many different individuals.

As Peterson notes, photographs can be too detailed, cluttered with too much unnecessary information. The same can be true of mathematical models. The point of modeling may not be to produce a detailed snapshot, but instead to understand. The goal might not be a thorough and realistic model for prediction, but a simpler model that enlightens or clarifies. Thus, when thinking about models, it is essential to keep in mind their purpose(s).

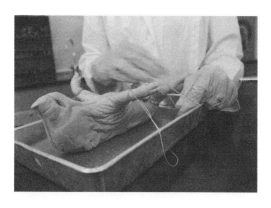

Figure 3.2. Left: A fetal pig (Figure source: Kelly Tippett/Shutter stock.com). Right: A mannequin (Figure source: Sandratsky Dmitriy/Shutterstock.com). Both are used to model the human body.

This point is made particularly vivid through the following example, from Blanchard, Devaney and Hall (2011, p. 2). Students of biology and medicine often perform dissections of a fetal pig as a way to learn about human physiology and anatomy. The pig, being a mammal like us, serves as a model for the human body. It might be better, I suppose, to dissect actual humans—and most medical students eventually do—but there are ethical issues associated with procuring and cutting up humans. So a model, the fetal pig, is used instead. Clothes designers, like doctors, also work with the human body. As they are designing clothes, they need to see what their work looks like when worn by a person. To do so, they make use of a different model of the human body—a mannequin.

So we have two different models for the human body, a fetal pig and a mannequin, a scenario illustrated in Fig. 3.2. Which model is better? The question is absurd. It is meaningless to critique a model without specifying the context—the goals or motivations for forming the model in the first place. Both are perfectly good models for what they are intended and would be disastrous if interchanged. I can't imagine wearing clothes designed for a pig, and I certainly wouldn't want to go to a doctor who had practiced surgery on a mannequin. Note that it is not a useful question to ask whether or not these models of a human are realistic. Neither are realistic, in that both leave out very important features of human beings. But that's the point—that's exactly what makes them effective models.

3.5 Levels of Models

The point of the previous section is to underscore that mathematics is harnessed in different ways when attempting to describe or understand the world. Models are constructed for different purposes: sometimes to predict, but other times to caricature or highlight particular aspects of a phenomenon without quantitative prediction in mind. This latter approach is sometimes referred to as *qualitative modeling* or qualitative mathematics. Models are also sometimes based on theoretical understanding, and other times are more empirical. For example, as previously mentioned, a model may aim to provide a faithful or mechanistic description of a phenomena. The point is not only to describe or even predict, but to make a statement of the way things really are. This type of modeling has its roots in physics, where we often begin an analysis with a precise knowledge of the behavior of a system's constituents and their interactions. We can thus write down rules that state how all constituents will behave—the *equations of motion*. Models of this sort are sometimes called *first-principles models*, because their starting point is a mechanistic, fundamental understanding of the entities that make up a system.

Another approach to modeling is *descriptive* or *empirical.* Here the goal is to capture or describe regularities present in data or a set of observations and not necessarily to elucidate the mechanism that produced the data. Much of the use of statistics is empirical in this way. One might want to know, for example, how strong a relationship there is between application of fertilizer and the productivity of tomato plants. One could carry out a linear regression and see if there is a statistically significant relationship between productivity and fertilizer usage. If such a relationship does exist, this analysis tells us nothing about why the relationship is the way it is—it just states that there is evidence in favor of there being a relationship.

Empirical models are used outside of statistics, as well. For example one might study a population of animals of some sort and observe that mortality as a function of population density is concave up—that is, the mortality increases faster than linearly with population density. This could arise because diseases spread more easily in a denser population than a less dense one. Or maybe there is some other explanation. All you know is the observed fact that mortality is concave up. A scientist building a model of this population might account account for this by adding a quadratic term to the death rate. The quadratic term does not get at the mechanism behind the mortality—indeed, the mechanism might not be known. But the term accounts for the empirical fact that the mortality increases faster than linearly as population increases.

So some models are mechanistic, beginning with a theoretical or fundamental understanding, while others are empirical, seeking simply to mimic some observed behavior or trend. And, of course, these categories are fluid. Many models combine elements of both. Also, the notion of fundamental is a relative one—what might be fundamental to a biologist might not be fundamental to a physicist. The notions of fundamental and empirical are not binary categories but exist along a context-dependent continuum.

Models also vary depending on the ontological status of their elements. Models are an abstraction—a simplification and an idealization of a process or system. The same can be said of the elements of a model, the terms or variable or objects from which the model is built. By ontological status, then, I mean the level or type of abstraction associated with the model elements. I'll give several examples of model elements at different levels of abstraction.

Much of physics is concerned with physical objects, like my water container, the chair I'm sitting on, or the toy mice that my cats like to play with. At the risk of drawing unwanted attention from philosophers, I will assert that these objects exist. I could use physics to determine how fast and at what angle to throw the toy mice for maximum distance. If I did so, I would treat the mouse as a point particle, a particle occupying an infinitely small point in space coinciding with the center of mass of the mouse. I would use the same projectile motion equations from introductory physics that apply to cannonballs or tennis balls or anything else. So using these equations entails an abstraction—I forget about the details of the toy mouse and treat it as a particle. You could do the same thing for a ball you are throwing to your dog, or a baseball thrown by the Red Sox rightfielder Mookie Betts to third base. This is, I would argue, not a very drastic abstraction. It is theoretically justified; the laws of physics assure us that a spatially extended object like a toy mouse really does behave as a particle positioned at the mouse's center of mass. Moreover, even though the idea of a particle is an abstraction, we are still talking about a real physical object that occupies a real point in space at a particular time.

In the previous chapter we focused on a differential equation that describes how the temperature of an object changes. Temperature is more of an abstraction than a particle. There are various ways of viewing or defining temperature: it is related to the average kinetic energy of the molecules that make up an object; it is a measure of the tendency of energy to flow in or out of an object (energy always flows from hot to cold); it is related to how the entropy of a system changes when its energy changes. All of these views of temperature can be shown to be equivalent. But regardless of which point of view one adopts, it seems fairly clear to me that temperature is a more abstract notion than position. Temperature can perhaps most concretely be thought of as a measure of average energy. The differential equation describing how a container of water cools off is then a statement about how an average changes, not how the energy of individual molecules change. The description is at a higher level—higher in the sense that it describes the situation in terms of coarse-grained or averaged variables.

More abstract still—or perhaps a different sort of abstract—is the notion of a population. We might write down a differential equation describing how a population grows. But a population is a coarse-grained description of what is really a collection of individuals. This becomes particularly clear when one models two populations, perhaps a predator-prey system, as we will do in Chapter 8. Such a model includes a term that accounts for the interaction between the two populations: the presence of prey leads to predator growth, and vice-versa. But this is a bit of a fiction. The populations don't interact, individuals do. An individual predator eats an individual prey. The point isn't that this sort of model is unjustified or wrong. All models are simplifications or idealizations of some sort, and I think it is important not to lose sight of this fact.

The notion of a density is also an abstraction. For example, one may describe the population of animals that are distributed in space by specifying their density. We might say that the density of elephants in a particular region of the savanna is 12 elephants per square kilometer. This is certainly a sensible thing to do, but it is not how I would experience elephants were I walking around the savanna. From my point of view, each point in space is either occupied by an elephant, or not. The density smooths over spatial variations, which may or may not be a good idea: it depends on the context of the model.

Scale matters too. One might also use a density to describe the concentration of smoke particles in the air. Perhaps there are some fires burning nearby while I am walking in the savanna, and the density of smoke particles is 25 micrograms per cubic meter. Just like the elephants, however, each point in space is either occupied

Figure 3.3. Some elephants in Tanzania. Both the elephants and the smoke in the air might be described with a density: elephants per square meter or smoke particles per cubic meter. (Image source: Hugh_Grant. Placed in the public domain (License: CC0 Public Domain. http://pixabay.com/en/baby-elephant-elephant-family-222 978/.))

by a smoke particle, or not. But unlike the elephants, I wouldn't experience the smoke this way—every breath I take would have very many smoke particles in it. Elephants and smoke particles operate at very different spatial scales: see Fig. 3.3. Whether or not describing a collection of objects (smoke particles or elephants) is a reasonable thing to do depends on the context. So the elements of a model are inevitably abstractions. Even familiar entities like temperature, population, and density are abstractions of a sort.

One response to these abstractions is to head in the other direction and to build a model concerned not with entire populations or densities, but instead focused on the individual agents or actors that make up a system. Models of this sort are known as *agent-based models* (ABMs) or sometimes, especially in ecology, *individual-based models*. In ABMs, rather than have a model describe how a population of elephants might grow, the model would consist of a representation of a number of elephants themselves, each of which would move, interact, reproduce, and die, according to some set of rules. The idea is to directly simulate the behaviors of individual elephants. This forces attention to the actual behaviors of the elephants, rather than thinking in average, coarse-grained terms as one would need to do when constructing a model at the population level.

Agent-based models have great appeal, because they dispense with a number of the abstractions that are inevitable when constructing models with differential equations. Personally, I find the case for ABMs compelling—what could be better than to work directly with the individual entities that make up a system? But in practice I have at times found it difficult to interpret results from an ABM. In the process of constructing an ABM one may need to introduce a great number of parameters, resulting in a model that is complicated and from which it is difficult to draw generalizations. Also, some of the fundamental choices encountered when constructing a model—is behavior stochastic or deterministic, rational or irrational, what forms do interactions take?—are not resolved by ABMs, but just pushed down to a lower level. Agent-based models are often viewed as a class of models all by themselves and are treated separately from other types of models. In my view this is unfortunate, as I think that ABMs are best seen as one among many different approaches to modeling. Moreover, there is not a bright line between agent-based and traditional models; again, I think things exist along a continuum.

3.6 Pluralistic View of Mathematical Models

The picture that I hope has emerged is that there is a diversity of styles and levels of mathematical models that are used in diverse ways for diverse purposes. Any model that has a temporal component can be viewed as a dynamical system. So there are thus many ways that dynamical systems are used across the sciences. A differential equation could be a first-principles model for the motion of an object as it falls due to gravity, or it could be a caricature-style model designed to hone intuition and lead to qualitative, and not quantitative, understanding. The mathematics of the differential equation is the same in both cases. But how we view it and what its purpose is might be very different.

What to make of this diversity of modeling styles and practices?[4] My intention certainly is not to deconstruct models to the point where there isn't any solid ground upon which to stand. Rather, I want to argue for a pluralistic middle ground. I have seen some who put a bit too much faith in their models, forgetting that what they are studying is just a model, and not the real thing. Others are dismissive of all models, arguing that they are misleading and not realistic. In my view neither of these extreme stances is justified.

What is the right way, then, to think about and assess a model? First, it is useful to remember that the abstractions that occur while formulating a model were done by choice. Models with their abstractions are human creations, not objective bits of reality, and so we might inquire about the goals of the model and the interests and intentions of those who made the model. Second, as Richard Levins writes, "Abstractions [and hence models] are not true or false but relatively enlightening or obscuring according to the problem under study. (Levins, 2006, p. 740)" Consider again the example of fetal pigs and mannequins as models for the human body. Both are enlightening if used for their intended purpose and obscuring if used for other purposes.

Different models elucidate different features of a phenomenon or process. Because of this, it seems a full understanding—if such a thing is even possible—will surely involve multiple models. Models at different levels and of different styles will shine light on

---

4. My thinking on this question has been strongly influenced by Richard Levins (2006) and several conversations I've had with him about issues of modeling and abstractions over the last few years.

different aspects of a problem while occluding others. A model usually is not the end of inquiry, but a chance to ask: what next? What have I left out of my model? As Levins puts it, "Where is the rest of the world? (2006, p. 745)." So I think that pluralism is called for, especially in the study of complex systems which, almost by definition, resist a simple description. Models of different styles and goals need not be viewed as contradictions of each other, but can be thought of as illuminating different facets of a complex crystal.

In the rest of this book, as we explore some of the central phenomena of chaos and dynamical systems, we will mostly use models designed for qualitative understanding. And as you encounter dynamical systems elsewhere, particularly in the study of complex systems, I think it is helpful to keep in mind that dynamical systems and mathematical models are used in many ways.

3.7 Further Reading

Much has been written about the theory and practice of using mathematical models to understand the world. A good well-referenced overview of epistemological and philosophical issues around the uses of models in science is Frigg and Hartmann (2012). A number papers written by biologists explore strategies and motivations for model building, including Levins (1966), May (2004), Levins (2006), Servedio et al. (2014), and Gunawardena (2014). The short paper by Epstein (2008) is a stirring and, in my opinion, effective defense of the role of models in the social and natural sciences. The essay by Healy (2016) is a provocative admonition against calls for detail or nuance in models. Good treatments of agent-based models include Ellner and Guckenheimer (2006, Chapter 8), Miller and Page (2007), Railsback and Grimm (2011), and Wilensky and Rand (2015).

4

CHAOS I: THE BUTTERFLY EFFECT

We are now ready to dive into one of the most interesting and important phenomena in dynamical systems. In this chapter I will introduce the phenomenon of chaos and the famous butterfly effect. The main example we'll consider is an iterated function known as the logistic equation. I'll introduce this dynamical system in the next section and then, in analyzing its behavior, we will be led into chaos.

4.1 The Logistic Equation

In this section I'll motivate the logistic equation. We will use this simple equation extensively in this and subsequent chapters as we encounter chaos. Our starting point is to consider population growth. For concreteness, let's imagine a particular population: the number of hippos in a lake somewhere. (See Fig. 4.1.) Given the population P_n this year—that is, the number of hippos today—what is the population next year, P_{n+1}? We are thinking of time as a discrete variable. We monitor the population from year to year, not continuously. So the population will be described by an iterated function, like those we looked at in Chapter 1.

$$P_{n+1} = f(P_n). \tag{4.1}$$

Figure 4.1. A pod of hippos. (Figure source: "Hippo pod edit" by Paul Maritz—Own work. Licensed under Creative Commons Attribution-Share Alike 3.0 via Wikimedia Commons. https://commons.wikimedia.org/wiki/File:Hippo_pod_edit.jpg.)

That is, number of hippos next year, P_{n+1} is a function of P_n, the number of hippos this year.

What would be a good choice for $f(P_n)$? How might next year's population depend on this year's? One possibility is that the population grows by a constant percent each year. This corresponds to the following iterated function:

$$P_{n+1} = rP_n, \qquad (4.2)$$

where r is the growth factor. In this model r is a *parameter*, a constant that can be varied to fit data or apply to different situations. If r is greater than one, the population will grow; if r is less than one, the population will decay.

Let's suppose that the hippo population increases by 10% every year. Then $r = 1.1$ in Eq. (4.2). An initial population P_0 of 1000 would lead to the following itinerary:

$$1000 \longrightarrow 1100 \longrightarrow 1210 \longrightarrow 1331 \longrightarrow 1464 \longrightarrow \cdots .$$
$$(4.3)$$

The population grows exponentially and will continue growing without bound. In some circumstances this can be a good model; there are indeed populations that grow exponentially. This might be the case if a small group of hippos moves into an otherwise hippo-less territory.[1]

However, in a finite world nothing can grow forever. As the hippos settle in to their new home there is initially a lot of food and room for everyone. But as the population grows, the limited food supply and/or overcrowding start to become an issue, and the growth will not continue. We seek a simple and general way to modify Eq. (4.2) so that the population does not grow indefinitely.

The following equation does the trick:

$$P_{n+1} = rP_n \left(1 - \frac{P_n}{D}\right). \tag{4.4}$$

In this equation r is again a measure of the growth rate and D is a quantity we can think of as a "doomsday" parameter. It is the population size at which the hippo overcrowding would become so severe that all the hippos would perish, leading to a population of zero. Equation (4.4) is known as the *logistic* equation; populations whose dynamics are given by Eq. (4.4) are said to exhibit *logistic growth*. In order to analyze Eq. (4.4), I'll need to choose values for the growth rate r and the doomsday population D. I'll use $r = 1.1$ and $D = 2000$.

So we now have two different growth equations: exponential, Eq. (4.2), and logistic, Eq. (4.4). Let's look at graphs of the right-hand sides of both of these equations, shown in Fig. 4.2. The exponential model, Eq. (4.2), is shown on the left of the figure. Note that no matter how many hippos there are this year,

1. One instance in which this has occurred is on the now-abandoned hacienda of a former drug lord in Columbia. An initial population of four hippos imported from Africa via New Orleans now may be as large as sixty (Kraul, 2006; El Espectador, 2014; Kremer, 2014; Howard, 2016).

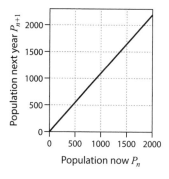 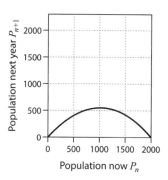

Figure 4.2. From left to right: graphs of Eqs. (4.2) and (4.4). In both equations the growth rate $r = 1.1$. In the right-hand equation, $D = 2000$.

there are always ten percent more next year. If there are 1000 hippos this year, there will be 1100 next year. If there are 2000 hippos, there will be 2200 next year. In contrast, Eq. (4.4) does not always lead to more hippos. If there are 1000 hippos this year there will be just over 500 next year. And if there are 2000 hippos this year, there will be no hippos next year. This justifies calling 2000 the doomsday population; it is the population at which the hippos are doomed.

We saw that the itinerary of $P_0 = 1000$ for Eq. (4.2) grows without bound; every year there are more hippos than the year before. What happens to an initial population of 1000 hippos using Eq. (4.4)? Will the population grow until it hits the doomsday number and disappears? Will the population decline gradually? Will it ever level off? Let's iterate and find out.

The itinerary for Eq. (4.4) with a seed of $P_0 = 1000$ is plotted with squares in Fig. 4.3. One can see that the population drops but levels off. It is approaching a fixed point at around 180.[2] Also in Fig. 4.3 I have plotted the orbits for two other initial conditions,

---

2. One can solve for the fixed points of this iterated function exactly by solving the equation $P = 1.1P(1 - \frac{P}{2000})$ for P. Doing so, one finds $P = 0$ and $P \approx 182$.

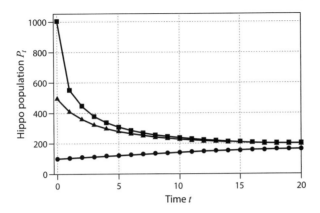

Figure 4.3. The time series for three different initial conditions (1000, 500, and 100), for Eq. (4.4) with $r = 1.1$ and $D = 1000$.

$P_0 = 500$ (triangles) and $P_0 = 100$ (circles). All orbits can be seen to approach the fixed point around 180. Evidently a population of 180 hippos is a stable equilibrium. Recall that our aim was to modify Eq. (4.2) so the population (hippos, or whatever), does not grow indefinitely. We were successful; Eq. (4.4) has done what we wanted.

As noted, Eq. (4.4) is known as the logistic equation. It is more common to write it in a slightly different and more general way. To arrive at this alternate formulation of the logistic equation, take Eq. (4.4) and divide both sides by D, to obtain:

$$\frac{P_{n+1}}{D} = r\frac{P_n}{D}\left(1 - \frac{P_n}{D}\right). \tag{4.5}$$

Next, define a new variable $x_n = \frac{P_n}{D}$. This new variable x_n is the population expressed as a fraction as the maximum possible, or "doomsday" population. So if $x = 0.4$, we interpret this not as four tenths of a hippo, but as a population that is four tenths of the maximum hippo population. With this new variable,

Eq. (4.5) is written as[3]:

$$x_{n+1} = rx_n(1 - x_n) \,. \tag{4.6}$$

This is the standard form of the *logistic equation*, a deceptively simple dynamical system that played a key role in the widening understanding and appreciation of chaos and dynamical systems in the 1970s and 80s. The paper by May (1976) was instrumental in drawing attention to and sparking interest in the logistic equation. (See also May (2002).) The logistic equation is a qualitative model of the sort described in Section 3.4; it is used to capture in broad strokes possible features of the dynamics of a population that has some limit to its growth. Next we begin an exploration of its properties.

4.2 Periodic Behavior

Let's explore the behavior of orbits of the logistic equation. To do so, I'll choose different values for r, make time series plots for each r value, and we'll see what we get. Before doing so, note that for the logistic equation in its standard form, Eq. (4.6), we are only interested in initial conditions between 0 and 1. Remember that

---

3. The procedure we just went through to define a new variable and arrive at Eq. (4.6) is an example of *non-dimensionalization*. In non-dimensionalization, one re-casts an equation in terms of dimensionless variables. In this example the variable P, which has units of hippos, was replaced with the variable x, which is dimensionless. The variable x has no units, being a ratio of two populations. Note that the dimensionless equation, Eq. (4.6) now has only one parameter, r. The original equation, Eq. (4.4) had two parameters: r and D. The process of non-dimensionalization has thus shown us that the logistic equation really has only one parameter, not two. The "doomsday" population level D can be made to have a value of 1 by rescaling P. This is equivalent to changing the units we're using to measure population.

Non-dimensionalization (also sometimes referred to as rescaling) is a useful and commonly used technique for simplifying iterated functions and differential equations. It is a bit of an art as well as a science. A clear discussion of non-dimensionalization is found in Otto and Day (2007, pp. 367–70). For other examples, see Ellner and Guckenheimer (2006, pp. 187–188), Edelstein-Keshet (2005, Section 4.5), and Strogatz (2001, pp. 64–66).

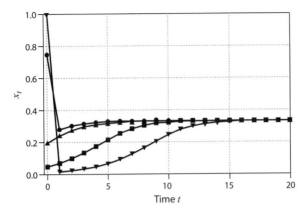

Figure 4.4. The time series for four different initial conditions for the logistic equation, Eq. (4.6) with $r = 1.5$. There is an attracting fixed point at $x \approx 0.33$.

we are measuring the population as a fraction of the "doomsday" value. So if $x = 1$, the population will die the next year, as you can verify by plugging $x = 1$ into Eq. (4.6). If $x > 1$, this means that there are more hippos (or whatever) than the doomsday population. The model breaks down at this point; the logistic equation gives a negative population value. So the moral of the story is that Eq. (4.6) only makes sense for x between 0 and 1.

I'll begin our exploration of the logistic equation by choosing a value of $r = 1.5$ for the growth parameter. In Fig. 4.4 I have plotted the time series for four different initial conditions: $0.05, 0.20, 0.75,$ and 0.99. We see that all orbits approach $x \approx 1/3$, and thus infer that this dynamical system has an attracting fixed point at $x = 1/3$. Any initial population between 0 and 1 will approach this equilibrium value of $x = 1/3$. And if the population is at $x = 1/3$ and is subject to some external influence that alters the population, the population will quickly return to $1/3$.

What about other r values? In Fig. 4.5 I have shown the time series for six different values of r. For all plots I have shown only one orbit, but the behavior I have shown is stable. Other initial conditions have the same long-term fate. In the upper

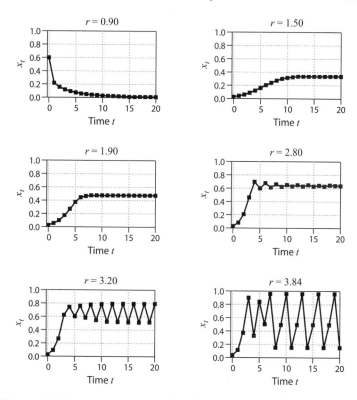

Figure 4.5. The time series Eq. (4.6) for six different r values. For all orbits the initial condition was $x_0 = 0.03$ except for the $r = 0.90$ plot which has an initial condition of $x_0 = 0.60$.

left plot, $r = 0.90$. This corresponds to a shrinking population. Accordingly, in this plot we see that the population decays and approaches zero. The upper right plot shows an orbit for $r = 1.5$. We have already considered this case in Fig. 4.4 and observed an attracting fixed point at $x = 1/3$. In the middle left plot the growth rate r is 1.90. We see an attracting fixed point at around $x = 0.45$, a bit larger than the stable population for $r = 1.50$. In the middle right plot, the growth rate is larger still: $r = 2.80$. The equilibrium value is also larger. There is an attracting fixed point at around $x = 0.64$. Now the population oscillates around this equilibrium value as it approaches it, rather than converging toward it from below as was the case for $r = 1.50$ and $r = 1.90$.

In the lower left plot of Fig. 4.5, where $r = 3.20$, we see periodic behavior. The orbit is pulled in to a period-two cycle; the population oscillates between approximately 0.80 and 0.51. This behavior is attracting; all initial conditions will end up getting pulled in to this period-two cycle.[4] Lastly, on the lower-right plot of Fig. 4.5 we again see periodic behavior, but this time the period is three. It takes three iterations for the population to repeat; the approximate population values in the cycle are 0.15, 0.49, and 0.96. This is stable; almost all initial conditions are pulled in to this period-three cycle.

To recap, we have seen two types of stable behavior: fixed points and cycles. We've seen cycles of two different periodicities: period two and period three. What is the nature of the transitions from one type of behavior to another as the parameter r is changed? Are there stable cycles of other periods? These questions will be addressed in Chapter 7. We'll now turn our attention to a non-periodic behavior exhibited by the logistic equation.

4.3 Aperiodic Behavior

Consider the logistic equation with $r = 4.0$:

$$f(x) = 4x(1-x). \tag{4.7}$$

As I did for other parameter values, I'll chose an initial condition, iterate, and make a time series plot. The result is shown in Fig. 4.6. Unlike our previous experiments, this time the orbit does not appear to be periodic. It looks like it might be about to become period three around $t = 7$, but then the population

---

4. Actually, there are some initial conditions that do not get pulled in to the attractor. However, if one chooses an initial condition at random, then with probability one that initial condition will approach the period-two cycle. A similar caveat applies for other r-values. See the discussion at the end of Section 1.5.

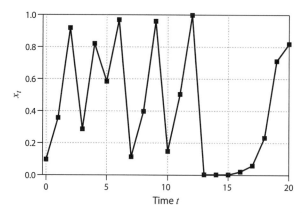

Figure 4.6. The time series for the logistic equation with $r = 4.0$. The initial condition is $x_0 = 0.1$. The orbit does not appear to be periodic.

crashes at $t = 13$ and grows again. What's going on? Perhaps it takes the population a while to settle into periodic behavior. Let's iterate longer and see what happens. In Fig. 4.7 I have plotted 300 iterates of the logistic equation with $r = 4.0$. Hmmm. It appears that the orbit is still not periodic. The population continues to bounce around fairly irregularly. At times it looks like the population is becoming periodic. For example around $t = 260$ it looks like period-two behavior appears, but one can see that this alleged periodic behavior does not persist.

It turns out that trajectories for the logistic equation with $r = 4.0$ are *aperiodic*—they never repeat. One could keep iterating forever, and the orbit would never return to exactly the same number. If you've not encountered this phenomenon before it is, I hope, a bit surprising and perhaps alarming. Iterating a function is a simple-minded and repetitive process. But for the function of Eq. (4.6), the result is perpetual novelty. Doing the same thing over and over again always results in something new. Several remarks are in order.

First, you may well be wondering how I am confident that the iterates never repeat. Figure 4.7 shows evidence that 300 iterates

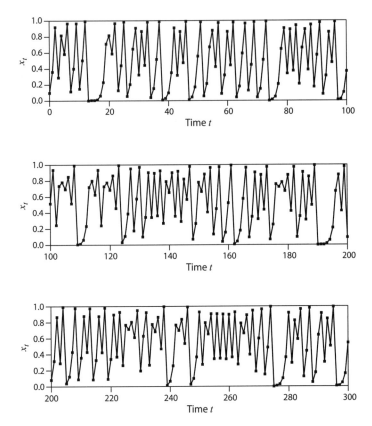

Figure 4.7. The time series for the logistic equation with $r = 4.0$. The initial condition is $x_0 = 0.1$. The orbit still does not appear to be periodic.

do not repeat. But it is a long way from 300 to infinity. The answer is that one can show mathematically that the orbits of the logistic equation, Eq. (4.6), are aperiodic. The techniques for doing so are a bit beyond the level of this book, but can be found in many textbooks on dynamical systems. See for example, Hirsch et al. (2004, Ch. 15) or Peitgen et al. (1992, Ch. 10).

Second, you might also wonder how it is possible for an itinerary to bounce around between 0 and 1 and never retrace its steps. Won't it run out of numbers and eventually have to start repeating itself? The answer, of course, is that there are an infinite number of numbers between 0 and 1, so new numbers will never be in short supply. But on a computer, numbers *will* run out. Even very large computers have a finite memory, and so can't keep track of an infinite number of digits or do arithmetic to infinite precision. An interesting discussion of this phenomenon can be found in Peitgen et al. (1992, Section 10.3). In practice, I don't think the finite precision and memory of computers is usually a concern, at least not for the applications covered in this book. But it is a good thing to be mindful of.

In any event, I think aperiodicity is an impressive feat for an iterated function. If I were to ask you to jump around a room not following any pattern and never landing on the same piece of floor twice, this would be a difficult task, as it would require remembering everywhere you had been previously. The logistic equation, however, has no memory. To compute the next value of the population, the function only requires as input the current population, not the full history of previous values. Unlike you jumping around the room, the dynamical system does not need to remember all of its previous positions. This is the first example of what will be a recurring theme throughout this book: complex or complicated behavior can emerge from a simple dynamical system.

Finally, suppose that we encountered a plot like Fig. 4.7 but did not know its origins. Perhaps it was population data gathered over many generations or the market price of some commodity. In such an instance we might be inclined to think whatever system we are studying is not following any rule—it is moving unpredictably, driven by random external influences. But Fig. 4.7 tells us that this need not be the case. A very simple, deterministic rule such as the logistic equation can yield results that appear to be rule-less. I think this is an important thing to keep in mind when studying complex systems and encountering seemingly random data. If a quantity is changing erratically we are not forced to conclude that it is subject to random external influences and is not governed by a rule or equation. It could be following a rule like the logistic equation with $r = 4.0$.

4.4 The Butterfly Effect

The time series plot shown in Fig. 4.7 seems random in that it does not appear to follow any pattern despite being generated by a simple iterated rule. In this section we will see that the orbits are not just seemingly random—they are also effectively unpredictable. Our starting point is Fig. 4.8. The top figure shows, on the same axes, two time series plots for the logistic equation. As before, the growth parameter $r = 4.0$. The orbits have different initial conditions. The orbit plotted with squares and a solid line has an initial condition of $x_0 = 0.10$. I will use y_t to refer to iterates for the other initial initial condition, plotted with circles and a dashed line. The initial condition for this itinerary is $y_0 = 0.11$.

The two orbits start off close together and for the first three or four iterates they remain fairly close. However, the two orbits are quite different by the fifth time step. Subsequently, the two orbits seem uncorrelated. They occasionally get close to each other again, as happens at $t = 14$. But the two trajectories quickly diverge. Another way to see this is to plot the difference between the orbits as a function of time. That is, plot $x_t - y_t$ versus t. Such a graph can be found in the bottom plot of Fig. 4.8. Initially the difference between the two orbits is small, and so the plot of $x_t - y_t$ is close to zero. As the two orbits pull apart from each other, $x_t - y_t$ moves away from zero. Sometimes $x_t - y_t$ is positive, and other times it is negative, because sometimes x_t is larger than y_t, and sometimes it's the other way around. The main thing to note is that quite quickly $x_t - y_t$ moves away from zero.

We see quite different behavior in a plot of two different trajectories for one of the r values we explored in the previous section. For example, in Fig. 4.9 I have plotted the itineraries for two

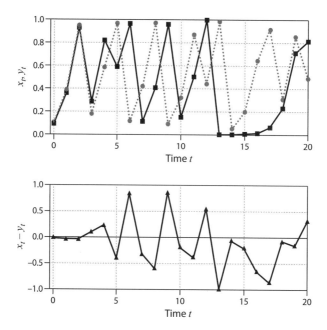

Figure 4.8. The top plot shows two time series for the logistic equation with $r = 4.0$. The initial conditions are $x_0 = 0.10$ (squares and solid lines) and $y_0 = 0.11$ (circles and dotted lines). The bottom plot shows $x_t - y_t$, the difference between the two time series.

different initial conditions for the parameter value $r = 2.8$. We saw in Fig. 4.5 that for this value of r the logistic equation has an attracting fixed point around 0.61. We see the same behavior in Fig. 4.9. In the bottom part of Fig. 4.9 we see that the difference between the two orbits approaches zero. This is just another way to see that the fixed point is attracting. Orbits that start far away get closer to each other as they approach the attracting fixed point.

Here is another way to think about what is going on in Fig. 4.8. Suppose that we have some physical or biological system that is exactly described by the iterated logistic function, Eq. (4.6). Let's say that the exact starting value is $x_0 = 0.1$, and so the exact trajectory of the system is given by the dark squares and solid lines

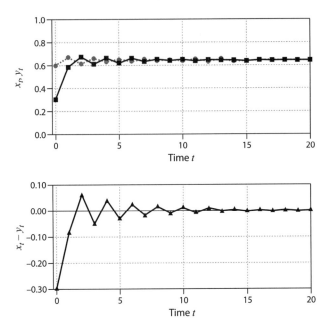

Figure 4.9. The top plot shows two time series for the logistic equation with $r = 2.8$. The initial conditions are $x_0 = 0.30$ (squares and solid lines) and $y_0 = 0.6$ (circles and dotted lines). The bottom plot shows $x_t - y_t$, the difference between the two time series.

in the upper plot on Fig. 4.8. We would like to predict the future behavior of the system. To do so, we make a measurement of the starting value. In this scenario we know exactly the rule governing the system, and since this rule is deterministic, we can predict with certainty the future behavior.

Or can we? There will inevitably be some error in our measurement of the initial condition. Suppose we measure 0.11 for the initial condition instead of the true value of 0.10. We would then use the slightly wrong initial condition, iterate it, and obtain a set of predictions for the future. These predictions are given by the circles and the dotted lines in Fig. 4.8. The lower plot in the figure can thus be viewed as the prediction error; it is the difference between the true values and the predicted values. We see that the prediction is only reasonable for a few time steps; by $t = 5$ the difference between the true value and our prediction is already 0.4.

In order to extend the duration of time for which our prediction is tolerable, we must increase the accuracy of our measurement of the initial condition. After all, this is the only source of error in this over-simplified scenario. We are assuming that the time evolution of the system is exactly described by the logistic equation. Suppose that we vastly improve our measurement techniques and now measure the initial condition to be 0.10001. This measurement is 1000 times more accurate than the previous one. What happens to our predictions?

In the top of Fig. 4.10 I have plotted the orbits for two initial conditions, 0.10001 and 0.1. As before, the exact values are plotted in dark squares with solid lines, and the prediction—the itinerary obtained by iterating the seed 0.10001—is plotted with circles and dotted lines. The two itineraries are right on top of each other until around $t = 14$, at which point they separate. This can also be seen in the bottom plot, which shows the difference between the two orbits—that is, the prediction error. We see that the prediction error is very small for the first dozen or so iterates, but then becomes quite large after $t = 15$.

With a more accurate initial measurement, we are able to accurately predict the orbit farther into the future. When our initial error was 0.01, as in Fig. 4.8, our prediction was good for around two time steps. In Fig. 4.10 our initial error is much smaller—only 0.00001. And we are able to predict out to $t = 14$. As expected, a more accurate initial measurement leads to a prediction that remains accurate for more time steps.

However, this improved prediction came at a high cost. We measured 1000 times more accurately but extended reliable prediction only by a factor of seven. This is rather disappointing. We have worked 1000 times harder and only benefited seven times more. This is an example of a phenomenon known as *sensitive*

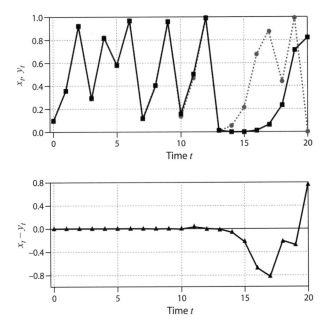

Figure 4.10. The top plot shows two time series for the logistic equation with $r = 4.0$. The initial conditions are $x_0 = 0.10$ (squares and solid lines) and $y_0 = 0.10001$ (circles and dotted lines). The bottom plot shows $x_t - y_t$, the difference between the two time series.

dependence on initial conditions. A dynamical system has sensitive dependence on initial conditions (SDIC), if very small initial errors grow and become arbitrarily large. I'll define SDIC more carefully in the following section, but for now we'll work with an informal definition: a dynamical system has SDIC if two initial conditions that start very close together eventually move very far apart.

Sensitive dependence on initial conditions is also known as the *butterfly effect*. The idea is that a system with SDIC is incredibly sensitive to any external perturbation. A tiny change in the state of a system, perhaps caused by the puff of air from the wings of a passing butterfly, could make a big difference in the system later on.

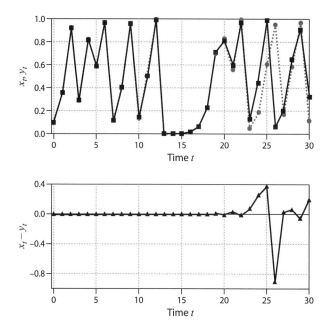

Figure 4.11. The top plot shows two time series for the logistic equation with $r = 4.0$. The initial conditions are $x_0 = 0.10$ (squares and solid lines) and $y_0 = 0.10000001$ (circles and light lines). The bottom plot shows $x_t = y_t$, the difference between the two time series. Note that the horizontal scale on this plot is larger than on the previous several plots.

The use of the butterfly metaphor is usually attributed to Edward Lorenz, one of the scientists responsible for the growth of the study of chaos and dynamical systems. For a fascinating short history of the term, see Hilborn (2004).

Let's look at one more example. Suppose now that we measure our initial condition incredibly accurately. We measure 0.10000001 and the exact value is 0.1. Our measurement is now accurate to one part in ten million. For how long will our prediction be accurate? Let's iterate and find out. In Fig. 4.11 I have plotted the itineraries for the initial conditions 0.10000001 and 0.1. The difference between the two orbits is shown in the bottom part of the figure. As expected, the two orbits are nearly identical for a while. However, by time $t = 24$, the itineraries have pulled apart and the measurement error becomes large.

In going from Fig 4.10 to 4.11 I have increased the measurement accuracy by a factor of 1000, but the time over which the prediction is good increases by less than a factor of two: from 14 to 24 iterates. This is SDIC again. Tiny differences between initial conditions grow and quite quickly become large. It is thus impossible to perform accurate long-term prediction of systems that have SDIC. I will discuss some of the implications of SDIC in the next chapter. I define SDIC more carefully and then define what it means for a dynamical system to be chaotic in the following section.

4.5 The Butterfly Effect Defined

In the previous section I described sensitive dependence on initial conditions—the butterfly effect—qualitatively. A system has SDIC if two orbits that have very similar initial conditions quickly become quite different. This qualitative definition gives the right idea. However, there is a technical definition for SDIC, and since SDIC is one of the central features of chaotic dynamical systems, I think it is worth stating the definition more carefully. I'll do so at several levels of mathematical abstraction. (That said, strictly speaking this section can be skipped, as I won't make reference to the formal definition of SDIC in subsequent chapters.)

Let's consider a function f on the unit interval.[5] This means that any input between 0 and 1 will remain between 0 and 1 if acted upon by f, as is the case for the logistic equation. Suppose we have an initial condition x_0. The function f has sensitive dependence on

---

5. This restriction to the unit interval is not necessary, but makes the definition that follows a bit simpler.

initial conditions if near to x_0 there is some other initial condition y_0 such that, when the two initial conditions are iterated with f, the orbits eventually get some specified distance away from each other. More formally: A function on the unit interval has sensitive dependence on initial conditions if for any δ, x_0, and $\epsilon \in [0, 1]$, there exists a natural number n and an initial condition $y_0 \in [0, 1]$ such that $|x_0 - y_0| < \epsilon$ and $|f^{(n)}(x_0) - f^{(n)}(y_0)| > \delta$. In words, this says that for any initial condition x_0, there is another initial condition y_0 that is within ϵ of x_0, such that the orbits of x_0 and y_0 eventually get further than δ apart.

It may help to think of this definition as describing a game, specified as follows. Somebody gives you values for δ, ϵ, and x_0. To be concrete, let's say that $\delta = 0.7$, $\epsilon = 0.04$, and $x_0 = 0.2$, and that the function is the logistic equation with $r = 4.0$: $f(x) = 4x(1 - x)$. Your task, then, is to find some initial condition y_0 that is within ϵ of $x_0 = 0.2$ whose orbit eventually gets farther away than $\delta = 0.7$ from the orbit of x_0.

With some experimentation, one can find a y_0 that meets this criteria. One such y_0 that works is $y_0 = 0.201$. It is within $\epsilon = 0.04$ of the other initial condition $x_0 = 0.2$, because $|x_0 - y_0| = |0.201 - 0.2| = 0.001 < \epsilon = 0.4$. The orbits of x_0 and y_0 are shown in the top plot in Fig. 4.12. The difference between the two orbits is plotted in the bottom figure. The solid horizontal lines on this figure are at ± 0.7. This makes it easy to see when the difference between the two orbits is first greater than $\delta = 0.7$. Evidently, this occurs at $t = 10$.

For a function to have SDIC, it needs to be possible to *always* succeed at tasks like this. That is, given *any* ϵ, *any* δ, and *any* x_0, it must always be possible to find a y_0 within ϵ of x_0 that eventually gets δ away. Note that there is no stipulation about how long it takes for the two orbits to separate by δ. It could take 10 iterations, as is the case in Fig. 4.12, or it could take $10,000$. In both cases, one would still say that the function has sensitive dependence on initial conditions.

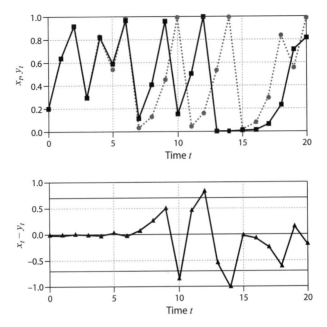

Figure 4.12. Two plots illustrating SDIC, as discussed in the text. The top plot shows two time series for the logistic equation with $r = 4.0$. The initial conditions are $x_0 = 0.2$ (squares and solid lines) and $y_0 = 0.201$ (circles and dotted lines). The lower figure shows the difference between the two orbits. The horizontal lines are at ± 0.7. We see that the magnitude of the difference between the two orbits becomes larger than 0.7 at $t = 10$.

I hope it seems plausible that the logistic equation with $r = 4.0$ has SDIC, especially if you've played around with plotting itineraries on your own. The picture that emerges from just a little bit of experimentation is that orbits get scrambled up very quickly; they bounce all over the place and no pair of orbits stays close together for long. This is the idea that is captured by the mathematical definition of SDIC laid out previously.

The fact that the logistic equation has SDIC for $r = 4$ can be proven rigorously—that is, using mathematical logic and without reference to computer experiments. Doing so is not tremendously difficult, but it is a bit of a long process that requires building up some mathematical definitions and constructions. A proof that the logistic equation has SDIC can be found in many mathematically-oriented dynamical systems texts such as Peitgen et al. (1992, Sections 10.5 and 10.6) and Alligood et al. (2006, Section 3.3).

4.6 Chaos Defined

I am now in a position to state what it means for a dynamical system to be chaotic. A dynamical system is *chaotic* if it has all of the following properties:

1. Its time evolution is given by a deterministic function.
2. Its orbits are bounded.
3. Its orbits have sensitive dependence on initial conditions.
4. Its orbits are aperiodic.

We have discussed the last two criteria, SDIC and aperiodicity, in the previous two sections. But we need to think a bit about the first two criteria. Let's consider the first criterion in the context of the logistic equation. The orbits for this dynamical system are most definitely governed by a deterministic function. The function is applied iteratively to the initial condition. There is no element of chance. The exact initial condition uniquely determines the orbit.

The second criterion stating that the orbits must be bounded is an important bit of mathematical fine print. If we omitted this criterion, then a very simple function would meet the definition for being chaotic. For example, consider the doubling function, $f(x) = 2x$. This function is deterministic, and its orbits are aperiodic. If one iterates this function using any non-zero seed, the result will never repeat. And the doubling function has sensitive dependence on initial conditions. Given any two non-identical initial conditions, the difference between the orbits doubles each time step and will eventually get larger than any threshold we set. Aperiodicity and SDIC are not noteworthy if the orbits are allowed to run toward infinity. They only become mathematically interesting if the orbits are bounded—that is confined to some finite interval.

So now you know the mathematical definition of chaos. A dynamical system is chaotic if it is deterministic with bounded, aperiodic orbits that have sensitive dependence on initial conditions. Students in my classes are sometimes a bit disappointed to learn that this is the official definition of chaos; they were expecting something weirder or stranger. I can understand their disappointment. The mathematical use of the word chaos does not align that well with its everyday use, where chaos denotes something that is lawless and completely without order or structure. Mathematical chaos is exquisitely lawful; a chaotic dynamical system obediently follows a deterministic rule in perpetuity. In fact, this obedience is the source of the butterfly effect. The system follows its rules so precisely that it depends sensitively on its initial conditions. To underscore this point, chaos is sometimes referred to as *deterministic chaos*. Regardless of whether or not it is the best name, chaos is here to stay. It has become an standard term in math and science.

There are a few other definitions of chaos, most of which are equivalent to the one just stated, which I believe to be the most common. A good discussion of the slight differences among definitions of chaos can be found in Bishop (2017). In some areas it is useful or interesting to draw distinctions among different sorts of chaotic behavior. There is a sequence of properties known as the *ergodic hierarchy* that distinguish among different degrees or strengths of chaos. The properties are, in order of increasing "chaos-ness": ergodic, weak mixing, strong mixing, Kolmogorov, and Bernoulli. Discussions of the ergodic hierarchy are often rather technical. A fairly accessible entry point is the essay by Frigg,

Berkovitz and Kronz (2014). In my experience the differences among these various types of chaos are interesting mathematics, but are often not relevant to specific applications of dynamical systems, nor to the big picture of the implications of the phenomenon of chaos.

4.7 Lyapunov Exponents

Sensitive dependence on initial conditions is an all-or-nothing deal. Either a dynamical system has SDIC, or it does not. But we could easily imagine that some dynamical systems are more sensitive than others. In this section I'll discuss Lyapunov exponents, a standard way to quantify the degree to which a dynamical system's behavior is sensitive to its initial conditions. Lyapunov exponents are quite commonly used, and they can be computed from data without knowledge of the function that generated the data. The material in this section will not be absolutely necessary elsewhere in this book, so you can skip or skim it if you want.

Lyapunov[6] exponents measure how the distance between orbits with two slightly different initial conditions grows (or shrinks) over time. To show how Lyapunov exponents are defined, I need to begin by fixing some notation. Let our two initial conditions be x_0 and y_0. The initial distance D_0 between the two orbits is given by

$$D_0 = |x_0 - y_0|, \qquad (4.8)$$

and let the difference between the orbits at time t be denoted by

$$D(t) = |x_t - y_t|. \qquad (4.9)$$

For an iterated function, $D(t)$ is defined only for integer values of t. For a differential equation, t is a continuous variable.

---

6. "Lyapunov" is also occasionally Romanized as "Liapunov" or "Ljapunov."

For a system with SDIC, we expect $D(t)$ to grow, since nearby orbits get pushed apart. Usually the growth is approximately exponential for small t, so we can write

$$D(t) \approx D_0 e^{\lambda t}, \tag{4.10}$$

where λ is known as the *Lyapunov exponent*. The intuition behind our expectation of exponential growth is as follows. Suppose that every time the function is iterated, the difference between two orbits grows by some fixed percentage. For example, if the growth was 5% each iteration, then after one iteration we would expect the separation to be

$$D(1) = 1.05 D_0, \tag{4.11}$$

where D_0 is the initial separation between the two orbits. And the separation after two iterations would be

$$D(2) = 1.05 D(1) = 1.05^2 D_0. \tag{4.12}$$

In general, the separation after n iterations is:

$$D(n) = 1.05^n D_0. \tag{4.13}$$

At every time step the initial separation between the two orbits, D_0, is multiplied by 1.05. Successive multiplication leads to exponential growth.

Equation (4.13) can be put into the form of Eq. (4.10) using logarithms:

$$D(n) = D_0 1.05^n = D_0 \left(e^{\ln(1.05)} \right)^n = D_0 e^{0.0488 n}, \tag{4.14}$$

since $\ln(1.05) \approx 0.0488$. (The second equality follows from the identity $e^{\ln(x)} = x$. The functions e^x and $\ln(x)$ are inverses of each other.) Thus, for this example the Lyapunov exponent λ is 0.0488. In this example the multiplication factor, 1.05, was the same every iteration. For a chaotic dynamical system this cannot be the case. In fact, sometimes the multiplication factor must be less than one—the difference between the orbits has to shrink occasionally.

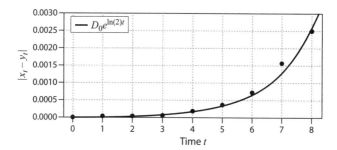

Figure 4.13. The circles show $D(t)$, the difference between two initially close seeds, for the logistic equation with $r = 4.0$. The line is the average growth rate of $D(t)$, given by: $D(t) = D_0 e^{\ln(2)t}$. The Lyapunov exponent is $\ln(2) \approx 0.693$.

If it didn't, the difference would grow continually and the orbits would not be bounded.

So for a dynamical system, the Lyapunov exponent gives a measure of the *average* exponential growth rate of two nearby orbits. A positive Lyapunov exponent means that the difference between nearby orbits grows, on average, exponentially with time. If the Lyapunov exponent is negative, this indicates that the difference between nearby orbits shrinks exponentially fast. The average exponential growth (or decay) rate is the given by the Lyapunov exponent.

The exponential divergence of nearby trajectories is illustrated in Fig. 4.13. Here I have used circles to plot the difference between two orbits with initial conditions of 0.17 and 0.17001. As usual, this is for the logistic equation with $r = 4$. For this value of r the Lyapunov exponent is known to be exactly $\lambda = \ln(2)$.[7] Thus, we would expect the average difference between two orbits for small t to be given by:

$$D(t) = D_0 e^{\ln(2)t}, \tag{4.15}$$

7. This is a standard result; see, e.g., Hilborn (2002, Section 9.5).

where D_0 is the difference between the two initial conditions. In Fig. 4.13 I have plotted Eq. (4.15) as a solid line. We can see that this is a fairly good approximation to $D(t)$.

We wouldn't expect Eq. (4.15) to be exact for any particular pair of orbits, because it represents average behavior. Also, as mentioned Eq. (4.10) holds only for small t. So care must be used if one wishes to use Eq. (4.10) to calculate the Lyapunov exponent, and a careful mathematical definition of Lyapunov exponents involves some subtleties. These issues are addressed in most intermediate and advanced texts on dynamical systems. A nice discussion of the calculation of Lyapunov exponents is Peitgen et al. (1992, Sections 10.1 and 12.5). See also Sprott (2003, Chapter 5) and, for a more mathematical treatment, Robinson (2012, Section 7.6).

Despite some mathematical complexities, the basic idea of a Lyapunov exponent is simple: it is the average exponential growth rate of differences between two orbits that start off very close. Equivalently, the Lyapunov exponent may be thought of as the average exponential rate at which prediction errors grow due to sensitive dependence on initial conditions. Geometrically, the Lyapunov exponent tells us the average rate of stretching per iteration that a typical orbit experiences. The larger the Lyapunov exponent, the greater the sensitivity on initial conditions. If a dynamical system has a negative Lyapunov exponent, it does not have SDIC, and hence is not chaotic. If a dynamical system has a Lyapunov exponent of zero, it may have SDIC or it may not; such systems are a borderline case. Dynamical systems that are chaotic but have a zero Lyapunov exponent are called *weakly chaotic*.

The definition of the Lyapunov exponent I have given applies to both iterated functions and differential equations. For higher-dimensional systems there is not just one Lyapunov exponent; instead, there is one exponent for each dimension. Lyapunov exponents can also be determined from data alone—that is, from a time series of data, without knowing the iterated function or differential equation that generated that data. References detailing techniques for doing so include Ott et al. (1994, Chapter 4) and Kantz and Schreiber (2004, Chapter 5).

This brings us to the end of this chapter. We will continue our discussion of chaos in the next chapter, where I'll present several ways of thinking about randomness that lead to the conclusion that the logistic equation, and other chaotic dynamical systems, are deterministic sources of randomness. I'll also offer some thoughts on the implications of sensitive dependence on initial conditions. Suggestions for further reading on chaos and the butterfly effect can be found at the end of Chapter 5.

5

CHAOS II: DETERMINISTIC RANDOMNESS

In this chapter we'll explore some of the philosophical and practical implications of chaos and the butterfly effect. The time series from the logistic equation appear random. Are they really? And how should one think about the butterfly effect? This chapter is a bit more philosophical and abstract the previous several chapters. The topics covered in Sections 5.1–5.3 are not needed in subsequent chapters, and so could be skipped or skimmed if one wishes.

In the previous chapter we have seen that time series produced by the logistic equation with $r = 4$ are highly irregular. For example, look again at Fig. 4.7. It is jarring how random the time series looks. How can something so erratic come from a simple, deterministic rule? The orbit can't really be random, can it? Actually, there are several points of view that lead to the conclusion that the time series generated by the logistic equation with $r = 4$ is indeed random. In this chapter I'll present two lines of reasoning that can lead to this conclusion. I'll begin by introducing symbolic dynamics, an alternate way of viewing and analyzing a dynamical system. We'll then use symbolic dynamics in Sections 5.2 and 5.3 to help us think about the randomness of the logistic equation.

5.1 Symbolic Dynamics

When one iterates the logistic equation one gets an itinerary consisting of a sequence of numbers between 0 and 1. (In this chapter all of the examples (with one clearly-noted exception) will be the logistic equation with $r = 4$. So at times I will simply refer to it as the logistic equation instead of the more cumbersome "logistic-equation-with-$r = 4$.") For example, iterating the seed 0.2, one obtains:

$$0.2, 0.64, 0.9162, 0.28901, 0.82194, 0.58542, 0.97081,$$

$$0.11334, \ldots . \quad (5.1)$$

We convert these numbers into discrete symbols in the following simple-minded way. If the iterate is less than 0.5, replace it with the symbol L, and if it is greater than or equal to 0.5, replace it with R. Applying this transformation to Eq. (5.1) yields

$$L, R, R, L, R, R, R, L, \ldots . \quad (5.2)$$

It seems as if the transformation I just did would surely have the effect of throwing out essential information, obliterating the dynamical system we are trying to learn about. It is as if I transform a sentence into symbols by using a C for every word that starts with a consonant and V for words beginning with a vowel. Applying this transformation to the previous sentence yields: VVVVVCVCVCCVVCCVCCCCVCVCCCCVC. If the original sentence is unknown, it cannot be inferred from the sequence of Cs and Vs. Even if I transformed the entire book this way, you would not be able to unambiguously decode the symbol sequence and recover the original words in the book.

However, for the logistic equation and the particular encoding L if $x_t < 0.5$, R otherwise, it turns out that, given an arbitrarily long sequence of Ls and Rs, it *would* be possible to go backwards. Suppose I choose an initial condition x_0, iterate it using the logistic equation to generate a time series, and then convert this into a symbol sequence of Ls and Rs. If I handed you this long symbol sequence, you would be able to figure out to reasonable accuracy what initial condition I used to generate the sequence, and you could then iterate this initial condition to determine the time series. This is not an obvious statement; it hinges on two facts. First, with each additional symbol one observes in the sequence one knows that the initial condition used to generate the sequence must have come from a smaller interval. Second, these intervals of potential initial conditions are non-overlapping; they partition the space of initial conditions.

Symbolic encodings that partition the space of initial conditions in this way are said to be *generating partitions*. One can now view the dynamical system as acting on the symbol sequences instead of the original numbers x between 0 and 1. A dynamical system of this type is known as *symbolic dynamics* or a *symbolic dynamical system*. A symbolic dynamical system obtained via a generating partition has the same topological properties as the original dynamical system: it has the same number of fixed points with the same stabilities. Symbolic dynamical systems are often much easier to study than the original dynamical system. Most exact results about the logistic equation, such as proofs of its aperiodicity and that it has SDIC, are obtained via symbolic dynamics.

Symbolic dynamics are great to work with, but a note of caution is in order. Symbolic dynamics only gives faithful information about the original dynamical system if the symbols are produced by a generating partition, and generating partitions are not always possible to obtain. In some situations there may be a partition that is not generating but which is nevertheless a useful or informative approximation, but this is not generically the case. In particular, a non-generating partition will make a dynamical system appear less random than it really is and may also introduce spurious complexity. As you might have guessed, there is a lot of elegant and (in my opinion) fun mathematics behind symbolic dynamics. Most intermediate to advanced books on dynamical systems with a mathematical focus discuss symbolic dynamics: for example Devaney (1989) and Robinson (2012). See also the text by Lind and Marcus (1995).

5.2 As Random as a Coin Toss

Let's return to the symbol sequence generated by the logistic equation and ask about its statistical properties. Do Ls and Rs appear equally often? I'll take an experimental approach to answering this question. I wrote a program to iterate the logistic equation one million times and record the fraction of Ls and Rs in the resulting symbol sequence. The results are:

$$\text{Fraction of L} = 0.501, \quad \text{Fraction of R} = 0.499. \qquad (5.3)$$

Evidently Ls and Rs occur with approximately equal frequency. What about sequences of pairs of symbols? How common are LL, LR, RL, and RR? Performing a similar experiment, I obtain:

$$\text{Fraction of LL} = 0.249, \quad \text{Fraction of LR} = 0.250,$$
$$\text{Fraction of RL} = 0.250, \quad \text{Fraction of RR} = 0.252. \qquad (5.4)$$

Again, we see that all pairs of sequences are approximately equally likely to occur. (The fractions do not add up to 1.0 because each individual fraction is rounded to three decimal places.)

You can probably see where this is headed. Let's do one more experiment with a million iterates of the logistic equation, monitoring the frequency of occurrence of trios of symbols. The results are:

$$\text{Fraction of LLL} = 0.124, \quad \text{Fraction of LLR} = 0.125,$$
$$\text{Fraction of LRL} = 0.124, \quad \text{Fraction of LRR} = 0.126,$$
$$\text{Fraction of RLL} = 0.125, \quad \text{Fraction of RLR} = 0.125,$$
$$\text{Fraction of RRL} = 0.126, \quad \text{Fraction of RRR} = 0.126. \qquad (5.5)$$

As before, we see that all triples of sequences occur with approximately the same frequency. For all these experiments I used a seed of $x_0 = 0.2$. If I used a different seed I would have gotten slightly different results.

If I generate longer and longer time series I would find that the frequencies become closer and closer to uniform. For example, all the frequencies in Eq. (5.5) would approach 0.125. This pattern holds as I consider longer and longer blocks of symbols. That is, I would find that all 16 possible sequences of four symbols (*LLLL*, *LLLR*, *LLRL*, and so on), would occur with frequency 1/16. Similarly, all 32 possible length-five sequences occur with frequency 1/32, and so on. Although I have presented these results via an experiment with a computer, they are established rigorously. Doing so involves showing that the logistic equation is equivalent to the shift map, a function whose associated symbolic dynamical system has the effect of shifting all symbols to the right by one. See Peitgen et al. (1992, Sections 10.5 and 16) or Hilborn (2002, Section 5.8).

Here is a way to think about what this all means. Suppose I generate two long symbol sequences. The first sequence I make with the logistic equation; I generate a very long orbit and then turn it into *L*s and *R*s. The other sequence I make by tossing a fair coin many times. Every time the coin comes up heads I write down *L*, and if the coin is tails I record a *R*. The result is a long sequence of *L*s and *R*s. Unlike the logistic equation and differential equations we have studied, tossing a coin is not a deterministic process. Making the same toss two times does not necessarily yield the same result. Functions with this property are called *stochastic*. A stochastic function incorporates an element of chance.

Suppose I handed you these two long sequences and didn't tell you which is which. You would not be able to tell the two sequences apart, even if you had infinitely long sequences. The two sequences are statistically identical. By this I mean that for both sequences the distribution of a single symbol is the same; both *L* and *R* are equally likely. The distribution of pairs of symbols are the same: *LL*, *LR*, *RL*, and *RR* are equally likely for both sequences. And the distribution of triples of symbols are the same, and so on.

Thus, the logistic equation has produced a sequence that is indistinguishable from a coin toss. I think this is an amazing fact. The logistic equation is, without any doubt, a deterministic dynamical system. And the process by which the time series is converted into symbols is deterministic as well. Yet the resulting sequence is indistinguishable from a coin toss, a paradigmatic example of a stochastic system. Can we really say that the logistic equation is producing random output? What could this even mean?

5.3 Deterministic Sources of Randomness

What do we mean when we say that something is random? One possibility is that something is random if it is generated by chance—it is the result of a stochastic process of some sort. But there is another approach to randomness that focuses on the sequences themselves and not necessarily the process that created them. In this section I will give a sketch of this approach, known as *algorithmic randomness*. In so doing, I'll closely follow the line of reasoning put forth by Flake (1999, Chapter 14).

The key idea is that a sequence is random if it is incompressible. This is perhaps best illustrated via some examples. Consider first the following sequence:

$$1000100010001000 \cdots . \qquad (5.6)$$

The "$\cdots$" indicates that the pattern continues indefinitely. Suppose you were asked to come up with a concise description of this sequence. You would probably say something like

$$\text{"The sequence is 1000 repeated forever."} \qquad (5.7)$$

By coming up with this short description, you have taken an infinite sequence, Eq. (5.6), and expressed it very concisely. You have thus *compressed* the sequence; Eq. (5.7) is much shorter than Eq. (5.6).

Achieving this compression was possible because there is a regularity or pattern to exploit. This regularity is what makes a shorter description possible. In fact, the shorter description essentially *is* a statement of the regularities in the sequence. Sequences that are compressible are not random. They have some regularities.

Let's consider another example, the sequence:

$$00111010110100011101011101\cdots. \qquad (5.8)$$

Unlike Eq. (5.6), this sequence does not have any obvious regularity. The "$\cdots$" indicates that the sequence continues in an irregular fashion. Suppose that you were tasked with finding a short description for the sequence shown in Eq. (5.8). What could you do? Since there are no regularities to exploit, all you can do is describe the sequence by restating it verbatim:

$$\text{"The sequence is : }00111010110100011101011101\cdots." \qquad (5.9)$$

The shortest description of the sequence is just the sequence itself. It has no regularities or patterns. Sequences which are their own shortest description are said to be *random* (at least in this algorithmic sense). This seems like a reasonable notion of randomness. Something is random if it has no regularities.

Let's now return to the logistic equation. The logistic equation generates symbol sequences that are statistically indistinguishable from the sequences generated by a tossing a fair coin. Such sequences are surely pattern-less and hence incompressible. Thus, sequences produced by logistic equation are indeed random.

But wait. This doesn't seem right. If the sequence was generated by the logistic equation, it can be described very compactly:

$$\text{Iterate } f(x), \text{ starting with the seed } x_0. \qquad (5.10)$$

$$\text{Convert to symbols : } 0 \text{ if } x_y \le 0.5, 1 \text{ if } > x_t. \qquad (5.11)$$

I might not be able to figure out what the seed x_0 should be, but I know that such a seed is out there, you might argue. We know there is such an x_0 because we just saw in the previous section that the logistic equation is as random as a coin toss. Thus, like a real coin, the logistic equation is capable of producing any and all sequences. Even if we can't find the right x_0, we know an x_0 exists, and so the sequence isn't really random.

All of this is true except for the last clause of the last sentence of the argument. Here's the catch. In order to come up with a short description of the sequence, we need to know the initial condition x_0. However, we need to know the initial condition *exactly*. The logistic equation has sensitive dependence on initial conditions. This means that to describe the infinitely long sequence, an approximate value for x_0 won't work.

So now we have shifted the problem. The issue becomes how to describe the infinitely precise initial condition. The overwhelming majority of initial conditions are irrational; if we choose an arbitrary initial condition, with probability one it will be irrational. And the vast majority of irrational numbers are incompressible; there is no pattern to their digits. Thus, the allegedly short description of Eq. (5.11) is actually not short at all. In order to describe the sequence via the logistic equation, our description must include the exact specification of the initial condition which will, with probability one, be incompressible. So sequences produced by the logistic equation really are algorithmically random.

Let's step back, collect some ideas and summarize. There are three concepts we've been considering:

1. **Deterministic.** A system is deterministic if it follows a well defined, unambiguous rule. The same input always gives the same output. If a system is deterministic, the itinerary is completely determined by knowledge of the rule and the exact initial condition.

2. **Stochastic.** A system is stochastic if there is an element of chance involved. The same input does not always yield the same output.

3. **Random.** A sequence is random if it is incompressible. A random sequence has no regularities or patterns.

Note that deterministic and stochastic are statements about the nature of the *process* that generated a sequence. In contrast, calling something random is a statement about the *outcome* of a process. An algorithmically random sequence can be generated by a deterministic or a stochastic process.

This is one of the key realizations to emerge from the study of dynamical systems: deterministic systems can produce results that are unpredictable (due to the butterfly effect), and are random (in the algorithmic sense). There is perhaps a tendency to think of randomness and determinism as opposites—that a rule-based system could not produce an unruly outcome. But we have seen that this is most definitely not the case. This is surely important to bear in mind when, as scientists, we encounter a seemingly random phenomenon.

Teaching these topics many times over the years I have found that some people are made uneasy by the notion of algorithmic randomness, and in particular the idea that deterministic systems produce randomness. To me this discomfort points out how entrenched the categories of random and deterministic are. Perhaps you don't like the definition of randomness as incompressibility. That's fine; some days I have mixed feelings about it, too. But regardless of ones feelings about algorithmic randomness, I think the main point of this section stands: deterministic systems can produce results that are unpredictable and statistically indistinguishable from a stochastic process.

As you have probably imagined, there is a lot of mathematical formalism needed to make rigorous the notion of incompressibility. The picture I have painted here is just the barest of sketches. To dig deeper, a good place to start is Downey and Reimann

(2007). See also Hutter (2007). Algorithmic randomness was introduced and formalized in the 1960s by Ray Solomonoff (1964*a*; 1964*b*), Andrei Kolmogorov (1965), and Gregory Chaitin (1966). An accessible discussion about distinctions between chance (loosely, stochasticity) and randomness is Eagle (2014).

5.4 Implications of the Butterfly Effect

What are the philosophical and scientific implications of the butterfly effect? There is not a single or simple answer to these questions. I think there are different lessons for different areas of study, and philosophers and scientists are certainly not in agreement about what chaos means. In this section I'll address this question in a number of different ways and will include some of references for readers who wish to read further.

Limits to Prediction in Practice

First, and most directly, the butterfly effect places strong limits to prediction. Prediction of a dynamical system that has sensitive dependence on initial conditions requires knowledge of the initial conditions to extreme accuracy. As a practical matter, this makes long-term prediction impossible. Small errors in measurement of the initial condition quickly become magnified by SDIC. Typically these errors grow exponentially fast. It is important to note, though, that chaotic dynamical systems *are* predictable—they're just not predictable for very long. How long they are predictable will vary, and in some cases the predictability window[1] might be sufficient for whatever task is at hand. Also, it is important to

---

1. The width of this predictability window (i.e., the time interval over which a predication is reliable) is related to the inverse of the Lyapunov exponent. The largest Lyapunov exponent λ, discussed in Section 4.7, is the average exponential rate at which two nearby trajectories separate. Thus, λ is the rate at which prediction errors grow. So $1/\lambda$ sets a predictability timescale. On average, the accuracy of a prediction will decrease by $1/e$ over a time interval of $1/\lambda$. See, e.g., Strogatz (2001, pp. 322–3).

remember that not all dynamical systems are chaotic. While chaos is a common phenomenon, non-chaotic (and hence predictable) dynamical systems are very common as well.

One response to the impossibility of long-term prediction is to give up trying to come up with a single best forecast—that is, choosing one initial condition, iterating, and reporting the resulting orbit as the prediction. The alternative is to instead start with a collection of initial conditions centered around our best-guess initial condition. We can then iterate the function or solve the differential equation, keeping track of all of the different solutions—one for each initial condition. The result will be a spread of trajectories that one can analyze to make statistical predictions about the dynamical system under study. The prediction method I just described is known as *ensemble prediction*, and the collection of initial conditions is referred to as an *ensemble*. Ensemble prediction is commonly used in meteorology—this is the origin of statements such as "there is a twenty percent chance of rain tomorrow." A non-technical discussion of ensemble prediction can be found in Smith (2007, Chapter 10). A technical review is Palmer et al. (2005).

Limits to Prediction in Theory?

I think the message of the butterfly effect is something qualitatively different than the simple observation that prediction is hard. Sensitive dependence on initial conditions imposes more severe constraints on prediction—constraints that are more fundamental than those implied by the simple observation that all measurements are imprecise and hence all predictions are imprecise. For example, let's go back to the illustration of the butterfly effect in Fig. 4.11. Here the two different initial conditions differed by a tiny amount. One initial condition was 0.1; the other was 0.00000001 larger. Let's imagine that rather than a population, this dynamical system described some physical quantity measured in meters. Then the first initial condition is 10 centimeters. The second initial condition is larger by just one nanometer. This is about the size of one glucose molecule or three times the size of a single carbon atom.

It seems to me that the difference between something that is 10 centimeters and 10 centimeters plus three carbon atoms is almost meaningless. It's not an issue of not having a good enough ruler—the difference between the two initial conditions seems physically inconsequential. As one looks at smaller and smaller lengths, we know from quantum mechanics that the world starts to "smear out," and objects cease to have well-defined positions. Quantum fluctuations will start to manifest themselves at or a bit below the nanometer scale. So the degree of accuracy needed to perform long-range predictions of a chaotic system is not just unattainable because of our coarse measuring instruments, but I would argue is not even theoretically possible.

Moreover, sensitive independence on initial conditions calls into question our ability to study a system in isolation. When we are studying a phenomenon in science there is often an implied boundary, and we assume that objects outside that boundary do not influence the phenomenon under study. For example, if I have students investigate the motion of projectiles in an introductory physics lab, I don't need to account for the gravitational pull of nearby Cadillac Mountain. However, the situation is different if we are studying a system that has sensitive dependence on initial conditions. In that case, we would need to keep track of all the butterflies in the vicinity, so to speak, since the trajectory of a chaotic system can be altered by tiny external forces. So it's hard to think of a chaotic system as being meaningfully isolated from its environment.

But this does not mean that it is impossible to study chaotic systems. Analyzing a chaotic systems involves calculating its Lyapunov exponents or understanding its attractor structure, as will be discussed in Chapter 9. These are features of a chaotic dynamical system that are stable, in the sense that they are robust to small perturbations and changes in initial conditions. So my students could carry out this type of analysis without worrying about the pull of nearby mountains.

Determinism, Laplace, and Newton

In Sections 3.2 and 3.3 I discussed the Newtonian worldview and Laplacian determinism. Briefly, the idea is that the world is mechanistic and deterministic. To Laplace's all-knowing demon, the future is an inevitable and predictable consequence of the present. Nobody thinks that Laplace's demon is an attainable goal, but arguably it stands as a distant beacon that science gradually moves toward. We seek to better understand the laws of nature, make better measurements, and harness more computational resources to unravel the consequences of these laws and measurements.

At one level, I think the phenomena of chaos need not obligate us to fundamentally alter the Newtonian worldview. The dynamical systems we've been studying are deterministic; the world is still one of cause and effect. Rules are in place. However, these rules don't lead to order or predictability in the conventional sense. Chaos doesn't say that the world is ruleless, but it tells us that those rules can lead to random results. We can still imagine that we live in a mechanistic universe, but the gears are not as straightforward as we had imagined.

In Section 3.2, I suggested that an additional aspect of the Newtonian worldview is the often implicit assumption that simple rules will lead to simple dynamical behavior. Without a doubt chaos tells us that this is not the case. We have already seen complicated behavior coming from the logistic equation—a very simple, non-linear rule. We will see additional examples of complicated and complex behavior arising from simple rules—both iterated functions and differential equations—throughout the rest of this book.

Some have suggested that the phenomenon of sensitive dependence on initial conditions frees us from the shackles of determinism and it gives a place for free will to exist again. I'm not so sure. Systems with SDIC are still deterministic even though the effects of that determinism are unpredictable. I think that this actually makes the question of free will even more problematic. The world is deterministic *and* unpredictable. How can free will be consistent with this state of affairs? Maybe this is why I'm not a philosopher. For more on chaos, determinism, and free will, see Bishop (2017, Section 7). See also the edited volume by Atmanspacher and Bishop (2014).

Order AND Disorder

Here is another way to think about how chaos might lead us to reconsider some aspects of the Newtonian worldview. In Fig. 5.1 are two time series generated by iterating the logistic equation, $f(x) = rx(1 - x)$. In the top plot the parameter r is 3.2, and we see an attracting cycle of period two. In the bottom plot, $r = 4.0$ and we see aperiodic, chaotic behavior. Suppose we didn't know that these plots were generated by the logistic equation, and instead we encountered these images after plotting some data. Perhaps the plots are populations of rabbits on different islands or commodity prices in different sectors of the economy.

Upon first looking at the two plots in Fig. 5.1 I suspect most would think they were very different. Moreover, it might not be clear how to think of the irregular time series. Perhaps one would attempt to describe it with a stochastic model; it would seem reasonable to suppose that it wasn't obeying any sort of rule at all. But as we know, both plots were generated by iterating a simple, deterministic equation.

Prior to the awareness of chaotic dynamical systems, I think most scientists would put the two plots in Fig. 5.1 in very different categories. These scientists would say that the world is made

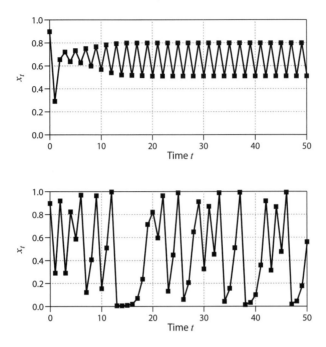

Figure 5.1. Time series generated by the logistic equation for two different *r* values. The top plot was made with $r = 3.4$, and the bottom plot used $r = 4.0$.

up of things that are orderly and things that are random, and we need different types of explanations for these different categories of phenomena. But chaos says that this isn't true. The same orderly, deterministic system can produce both predictable periodic behavior and unpredictable chaotic behavior. So randomness and order aren't completely separate things that we need to think of in different ways. In a sense, they are two sides of the same coin.

5.5 Further Reading

Chaos is, of course, covered in all textbooks on chaotic dynamical systems. There is also ample discussion of chaos and the butterfly effect in popular science writing. In addition to Gleick (1987) and

Stewart (2002), I would also recommend the engaging and accessible book by Ruelle (1993). The philosophers of science Kellert (1993) and Smith (1998) have both written books in which they address a range of philosophical and scientific implications of chaos. Kellert's views are closer to my own, but both books are excellent and highly recommended. The Stanford Encyclopedia of Philosophy entries on Causal Determinism by Hoefer (2016) and Chaos by Bishop (2017) would be good places to start for an overview of some of the issues raised in this chapter and pointers to further references. Additional discussions of the relations among chaos, determinism, indeterminism, and (un)predictability, can be found in Hobbs (1991) and Bishop (2008). A clear, brief treatment of algorithmic randomness and chaos is the thought-provoking and influential review article by Ford (1983). The review article by Crutchfield, et al. (1986) is also recommended.

6

BIFURCATIONS: SUDDEN TRANSITIONS

In this chapter we will leave the world of chaos and turn our attention to another interesting phenomenon exhibited by dynamical systems: bifurcations. As we shall see, bifurcations are sudden qualitative changes in the behavior of a system as a parameter is varied continuously. Bifurcations have not captured the popular imagination nor gotten as much hype as chaos, but they are an important and surprising phenomena that are definitely worth understanding.

I will begin in Section 6.1 by introducing a differential equation version of the logistic equation that we studied in Chapter 4. In Section 6.2 I will modify this equation by adding a harvest term. This will lead us to bifurcations and bifurcation diagrams in Section 6.3. In Section 6.4 I discuss bifurcations more generally, and in Section 6.5 I will comment on two notions related to bifurcations: catastrophes and tipping points. Finally, in Section 6.6 I discuss hysteresis, a phenomenon that endows systems with a type of memory or path dependence.

6.1 Logistic Differential Equation

Our starting point for learning about bifurcations is the logistic differential equation, which is the differential equation version of the discrete logistic function that served as our central example in

Chapters 4 and 5. Recall that the logistic equation is given by:

$$x_{n+1} = rx_n(1 - x_n) . \tag{6.1}$$

I motivated this equation by arguing in Section 4.1 that it represented a simple model of a population in which there is a limit to growth. The differential equation version of the logistic equation is:

$$\frac{dP}{dt} = rP\left(1 - \frac{P}{K}\right) . \tag{6.2}$$

This is a differential equation similar to Newton's law of cooling, which was our initial example of a differential equation, introduced in Chapter 2. We will use a slightly modified version of Eq. (6.2) to explore bifurcations. Does a small change in the differential equation always lead to a small change in the long-term behavior of the solutions? (Spoiler alert: no.)

As with the iterated logistic equation, the logistic differential equation, Eq. (6.2), describes the growth of a population. For concreteness sake, let's imagine that this equation describes the population of fish in a lake or a bay. The variable P is the population, measured in total fish number. Equation (6.2) tells us how dP/dt, the growth rate of P, depends on P. The quantity r is a parameter that changes the growth rate of the population. The quantity K is, for reasons we shall see shortly, known as the *carrying capacity* of the system.

Let's analyze this equation using the techniques developed in Chapter 2. In the top part of Fig. 6.1 I have plotted the right-hand side of Eq. (6.2). For this plot and all subsequent plots in this chapter I have chosen $r = 3$ and $K = 100$. The equation is an upside-down parabola, crossing the horizontal axis at 0 and 100. When the fish population is between 0 and 100, dP/dt is positive, so the population is increasing. When the population is larger than 100, we see that the rate of change of the population dP/dt is negative, so P decreases. There is thus an attracting equilibrium at 100 and a repelling equilibrium at 0.

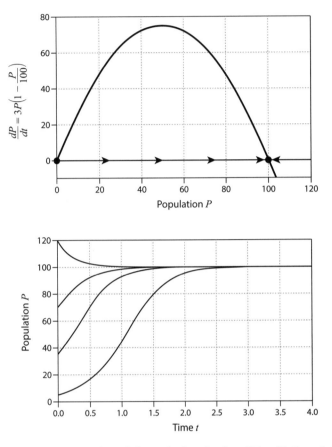

Figure 6.1. Top: A plot of the right-hand side of Eq. (6.2) and its phase line. There is a stable fixed point at $P = 100$ and an unstable fixed point at $P = 0$. Bottom: The solutions $P(t)$ to Eq. (6.2) for four different initial conditions: $5, 35, 70,$ and 120. All solutions approach the stable equilibrium at $P = 100$.

So we expect to see a stable population at 100. I have chosen the parameters r and K in Eq. (6.2) to be numerically convenient, not biologically realistic. The goal of the next several sections is to illustrate the phenomenon of bifurcations, not faithfully model actual fisheries.

On the top plot of Fig. 6.1 I have drawn the phase line for this differential equation. The two fixed points are indicated as solid circles. Note that the arrows point to the right, indicating population growth, when dP/dt is positive, and point to the left, indicating population decline, when dP/dt is negative. The lower plot in Fig. 6.1 shows solutions $P(t)$ to the differential equation. I have shown the solutions for four different initial populations: 5, 35, 70, and 120. The solution's behavior is consistent with that illustrated in the phase line: all solutions are pulled toward the stable equilibrium at 100.

The quantity K—in this example 100—is known as the *carrying capacity*. The carrying capacity is the stable equilibrium for the population. Ecologically, the idea of a carrying capacity should be taken with a grain of salt. Some ecologists feel that its use implies that there is an underlying stability to populations, which is not warranted empirically. These critiques need not concern us here. (If you want to read more about carrying capacity, an interesting short history of the term, in ecology and elsewhere, is Sayre (2008).) The main point of the logistic differential equation in our context is to serve as a simple model of a population in which there is some limit to growth. And the point of this particular example is to illustrate bifurcations, which I will do in the next section.

6.2 Logistic Equation with Harvest

We will now modify the logistic differential equation to account for the effects of fishing. Doing so will add an additional parameter to the equation. We will see that small changes in this new parameter lead to large, sudden changes in the system's behavior.

One way to modify Eq. (6.2) to account for fishing is as follows:

$$\frac{dP}{dt} = rP\left(1 - \frac{P}{K}\right) - h. \tag{6.3}$$

Fishing is accounted for by the $-h$ term. It has the effect of removing h fish per unit time, decreasing the growth rate by h. This is often called a harvest term—hence the letter h—and Eq. (6.3) is often called the logistic equation with harvest.

What is the effect of fishing? Presumably fishing leads to fewer fish. But to what extent? And what happens as the fishing increases—as h gets larger and larger? Let's analyze the equation and see. We begin by considering the case where $h = 25$. (As before, $r = 3$ and $K = 100$.) A plot of the right-hand side of Eq. (6.3) for these parameter values is shown in the upper right of Fig. 6.2. The upper left plot in the figure shows the $h = 0$ case, which we analyzed in Fig. 6.1. The logistic equation without harvest, $h = 0$, is shown in the upper left of the figure. Let's compare the two plots.

Subtracting h from a function shifts it down on a graph by h units. So the upper-right curve in Fig. 6.2 is 25 units lower than the upper-left curve. What happens to the equilibria when we allow fishing at a rate of $h = 25$? We can see on the figure that the two equilibria get closer to each other. The stable equilibrium that was formerly at $P = 100$ now appears to be at $P \approx 90$. And the unstable equilibrium that was formerly at $P = 0$ now has shifted up to $P \approx 10$. A fish population that is less than 10 will decay and die out.[1]

---

1. Exact values for the equilibria can be found with a bit of algebra. The equilibrium condition is that the rate of change is zero: $dP/dt = 0$. Using the values $K = 100$ and $r = 3$ in Eq. (6.3), this condition is:

$$3P\left(1 - \frac{P}{100}\right) - h = 0 . \tag{6.4}$$

Solving for P, one finds

$$P = 50 \pm 10\sqrt{25 - \frac{h}{3}} . \tag{6.5}$$

As long as $h < 75$ the argument of the square root is positive there are two real solutions, symmetric about $P = 50$, as we can see in Fig. 6.2.

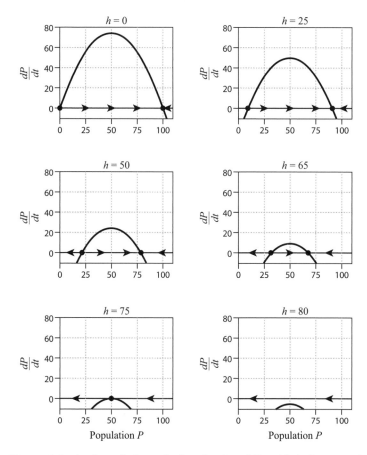

Figure 6.2. A plot of the right-hand side of Eq. (6.3) for several different harvest rates h. Also shown is the phase line for each h value.

What happens if we increase the fishing rate further? The middle-left plot on Fig. 6.2 shows the right-hand side of Eq. (6.3) for $h = 50$. As expected, the graph is shifted further down. This has the effect of moving the two equilibrium points even closer together. The stable equilibrium is now at $P \approx 79$. This is not surprising; more fishing should lead to less fish. We've seen that increasing the fishing rate from 0 to 25 to 50 decreases the equilibrium population from 100 to 90 to 79.

Continuing to increase the fishing rate h will cause the equilibrium population to decrease further still. In the middle-right plot in Fig. 6.2 the fishing rate h is 65, and the curve is shifted even lower. The stable equilibrium is now at around $P = 68$. The unstable equilibrium is at $P \approx 32$. If the fish population ever dips below 32, it will decrease and die out.

Suppose the fishing rate was exactly 75. This scenario is shown in the bottom-left plot on Fig. 6.2. The curve just barely touches the horizontal axis. There is thus only one equilibrium now, $P = 50$. This equilibrium is stable from the right, but not the left. In other words, if the population is at 50 and increases by a few, it will return back to $P = 50$. However, if the population is 50 and it decreases by a few, the population will continue to decrease and all the fish will die. (Mathematically, equilibria of this sort are sometimes called *semi-stable*.)

Finally, in the bottom-right plot in Fig. 6.2 I have shown the situation for $h = 80$. Here the fishing rate is large enough that the entire curve is below the horizontal axis. This means that for all population values P the growth rate is negative. Regardless of the current number of fish, if $h = 80$, the population is doomed.

Figure 6.2 shows us how the behavior of Eq. (6.3) changes as the parameter h is changed. We see that the system transitions from having two equilibria to zero. If h is less than 75, the system has two equilibria (one stable and one unstable), and if h is greater than 75, the system has no equilibria. This transition is known as a *bifurcation*. I will define bifurcations more carefully, and also discuss what this means for fish—and for the study of complex systems. First, however, I will introduce bifurcation diagrams—a type of plot that is a standard way to view bifurcations. Bifurcation diagrams are a bit abstract, so I'll construct them in several steps. The discussion that follows might be a bit pedantic, but my experience has been that being careful and almost tedious is the best way to help people understand bifurcation diagrams.

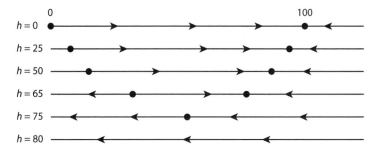

Figure 6.3. The phase lines for Eq. (6.2) for the six h values shown in Fig. 6.2: 0, 25, 50, 65, 75, and 80.

6.3 Bifurcations and Bifurcation Diagrams

We have been exploring how the behavior of the differential equation, Eq. (6.3), changes as the harvest rate h is varied. Let's see if we can view this more comprehensively than we did in Fig. 6.2, which shows the behavior for only six h values. It would be nice to be able to display the system's behavior for all h values, not just these six. To do so, let's start with the six phase lines shown in Fig. 6.2 and imagine cutting them out and stacking them up, as shown in Fig 6.3. Note that at $h = 0$, when there is no fishing, there are two equilibria, a stable one at $P = 0$ and an unstable one at $P = 100$. As h is increased, which corresponds to moving down in Fig. 6.3, the two equilibria get closer together. At $h = 75$ the two points momentarily merge into one. Then at $h = 80$ there are no equilibria. The fish population will die off regardless of the initial population.

To more clearly observe this behavior, I will take the phase lines in Fig. 6.3 and rotate them 90 degrees counter-clockwise. I will construct a graph on which the harvest rate h is the horizontal axis and I will space the phase lines accordingly. The result of doing this is shown in Fig. 6.4. Please take a moment and make sure you understand how to go from Fig. 6.3 to Fig. 6.4.

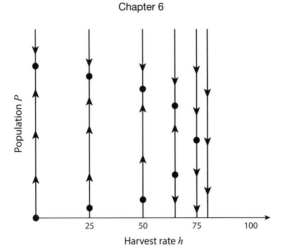

Figure 6.4. The same phase lines shown in Fig. 6.3 rotated ninety degrees counter-clockwise and arranged to-scale according to their *h* values.

Figure 6.4 shows us how the qualitative behavior of the differential equation, as shown by the phase line, changes as the harvest rate *h* changes. Our last step is to "connect the dots"—to plot many more phase lines so we can see the value of the equilibria for all *h* values, not just the six I have shown in Fig. 6.4. Doing so yields Fig. 6.5, a type of plot known as a *bifurcation diagram*. In the figure the stable equilibrium is shown as a solid line. It starts at 100 for zero fishing (*h* = 0), and decreases as *h* is increased. The unstable equilibrium is the gray line. It is also common to use dashed lines for unstable equilibria. We see that the unstable equilibrium rises as the harvest rate *h* increases. The two equilibria meet at *h* = 75. For *h* larger than 75 all initial conditions decrease and there are no equilibria. (In Fig. 6.5 the lower limit for the population *P* is −10. Negative population values are not meaningful biologically, but I included them on the plot so there was room for all the arrows.)

Bifurcation diagrams such as Fig. 6.5 are fairly abstract constructions—it is a collection of phase lines plotted as a function

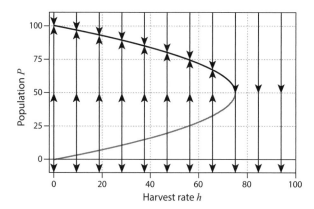

Figure 6.5. The bifurcation diagram for the logistic differential equation with harvest, Eq. (6.3). The black curve shows the stable equilibria while the gray curve, the unstable equilibria.

of a parameter. If you get confused when thinking about or working with bifurcation diagrams, I suggest going back to this example and remembering that a bifurcation diagram is nothing more than a bunch of phase lines "glued" together. With a bifurcation diagram in hand, we can figure out the phase line, and hence the qualitative dynamics, for any parameter value we choose. We treat the bifurcation diagram as a dictionary of phase lines: choose the h we're interested in, look it up on the bifurcation diagram, and read off the phase line. For example, looking at Fig. 6.5 we see that at $h = 40$ there is an attracting fixed point a bit above 80 and a repelling fixed point a bit below 20. At $h = 90$, we see that there are no fixed points and all P values decrease continually.

So now we know how to make a bifurcation diagram, but what are they used for? And what is a bifurcation, anyway? A *bifurcation* is a sudden change in the qualitative behavior of a dynamical system as a parameter is varied continuously. Fig. 6.5 shows that the logistic equation with harvest, Eq. (6.3) has a bifurcation at $h = 75$. Below $h = 75$ the dynamical system has two equilibria, one stable and one unstable. Above $h = 75$ there are no equilibria. This is what is meant by a change in the qualitative behavior of a dynamical system—the number of fixed points or their stability changes. In contrast, there is not a bifurcation at $h = 40$. Below and above $h = 40$ the qualitative behavior is the same: there is one stable and one unstable equilibria. The position of these equilibria change slightly, but this is not considered a qualitative change in the dynamical system's behavior.

The existence of bifurcations has some important and interesting implications. Let's think about what the bifurcation diagram in Fig. 6.5 is telling us. If there is no fishing, $h = 0$, we know that there is a stable equilibrium at $P = 100$. If we allow fishing, the stable population decreases. For example, if $h = 25$, the stable population decreases to $P \approx 90$, as we have seen. This certainly makes sense; if we introduce fishing, the fish population goes down. If we increase the fishing a little bit, the population decreases a little bit. No big deal. We could keep going in this manner: increase the fishing rate a little bit every year, and each year the fish population would decrease a little bit. This story will come to an abrupt end, however, at $h = 75$. When we cross the bifurcation, going from h a bit less than 75 to h a bit more than 75, the fish population drops suddenly from 50 to 0. There are no stable values in between. The fish population crashes.

Once h is greater than 75, it might take a year or two for the crashing fish population to become evident. During this time the fish population would be declining all the while. We could imagine that it is not until the population reaches 25 that some action is taken. Perhaps it is then decided to enact legislation so that $h = 60$. This would seem like a safe fishing level, since previously when the fishing rate was 60 there was a stable equilibrium a bit below $P = 70$. However, if $h = 60$ and $P = 25$, the fish are still in trouble. A population of $P = 25$ is below the unstable equilibrium at $P \approx 27$. The result is that the population continues to decline.

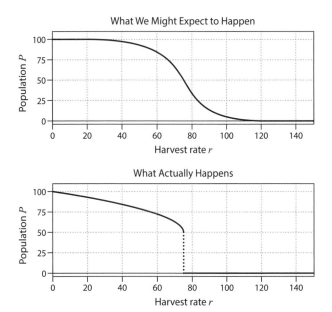

Figure 6.6. Two scenarios for how the fish population decreases as the fishing rate h is increased. In the top plot, the population decreases smoothly and continuously as h increases. In the bottom plot, there is a bifurcation at $h = 75$. The equilibrium population jumps from 50 to 0 at $h = 75$.

Here is another way to visualize and think about what is going on with the fish. We might expect that a small increase in fishing will always lead to a small decrease in the number of fish. As the fishing rate h is increased, the number of fish might not fall off linearly, but we would expect a smooth decrease until eventually there are no fish left. This is the scenario depicted in the top plot of Fig. 6.6. But what actually happens in this model is shown in the lower plot on Fig. 6.6. The equilibrium population P decreases smoothly and continuously for a while, but then at $h = 75$ there is a sudden transition. The equilibrium population jumps from 50 to 0.

This is potentially bad news, especially in a realistic setting in which it would not be possible to accurately measure the fish population directly. Instead, one would have to rely on an indirect measure, perhaps how hard it is to catch a certain amount of fish. More crucially, we would not know the parameters of the model exactly, and so we would not know exactly where the bifurcation point is. And as we have seen, the intuition that a small change in the fishing rate h leads to a small change in the fish population P is not always correct.

The preceding analysis focused on the logistic differential equation with harvest, Eq. (6.3). But what about the collapse of real fisheries, as opposed to mathematical models? And what about other phenomena in which abrupt changes occur? It is unarguable that physical, natural, and social systems sometimes undergo sudden changes (see, e.g., Scheffer (2009)). Fisheries and economies collapse, climates change quickly (on geological time scales) from moist to dry, evolution often consists of periods of sudden change, and social norms and practices sometimes change at surprising times with surprising speed. The world consists of behavior like both of the plots in Fig. 6.6.

Bifurcations of the sort we have examined here surely do not explain all of these phenomena. But studying dynamical systems does demonstrate that sudden transitions can occur in very simple models. Moreover, these are models that are themselves continuous. The right hand side of the logistic equation with harvest, Eq. (6.3) is continuous and smooth. It is perhaps surprising—and in my opinion interesting and fun—that one can get discontinuous behavior from a continuous model. Like a lot of things in math, I think bifurcations are kind of obvious or at least not very deep once one encounters them. But until one has seen an example of a bifurcation, it might be counter-intuitive to think that a population described by a smooth and continuous differential equation could undergo a sudden change.

6.4 General Remarks on Bifurcations

So far in this chapter I have focused on a single example of a bifurcation—the disappearance of the stable and unstable equilibria that occurs in the logistic equation with harvest. This section contains some more general remarks about bifurcations.

First, it turns out that while bifurcations are common, there are only a few different types of them. The reason for this is as follows. Bifurcations occur when the number or stability of equilibria change. For differential equations of the form $dP/dt = f(P)$, equilibria occur at the populations P at which $f(P) = 0$. (Of course the differential equation need not describe a population, but I'll continue to use this language for the sake of concreteness.) Geometrically, equilibria occur when the function $f(P)$ crosses the horizontal axis. Bifurcations occur when there is a change in the number of times $f(P)$ crosses the axis.[2] There are only a handful of ways that the number of times $f(P)$ crosses the axis can change as a parameter is varied, assuming that $f(P)$ depends continuously on whatever parameter is being varied. One such way that the number of crossings of $f(P)$ can change was illustrated in Fig. 6.2. This gave rise to the bifurcation diagram shown in Fig. 6.5. We will see a different example in the subsequent section.

So bifurcations fall neatly into several classes. For example, any differential equation in which two equilibria disappear by shifting $f(P)$ downward as a parameter increases—as in Fig. 6.2—belongs to the same class. Bifurcations in this class are generally

---

2. There also is a bifurcation if there is a change not in the number of times $f(P)$ crosses the horizontal axis, but in how that crossing occurs. If $f(P)$ crosses the axis from below, that equilibrium point is unstable; if it crosses from above, the equilibrium point is stable. There would be a bifurcation if there is a change such that $f(P)$ at an equilibrium point switches from crossing below to above, or vice-versa.

known as *saddle-node bifurcations*. There is a creative and not-quite-standard terminology used to describe bifurcations. The saddle-node bifurcation is sometimes referred to as a *fold*, *tangent*, *turning-point*, or *blue-sky* bifurcation. There are other classes of bifurcations, including *transcritical bifurcations* and *pitchfork bifurcations*, the latter of which comes in two flavors, *supercritical* and *subcritical*. To be honest, except for "pitchfork bifurcations," which actually resemble pitchforks, I find the names for bifurcations to be unhelpful. There is a standard example for each type of a bifurcation known as the *normal form*. A normal form for a bifurcation can be thought of as a prototype for all bifurcations in each class.[3] More complex dynamical systems will have the same bifurcations as simpler ones. Thus, we can learn about bifurcations exhibited by complex models by studying simple ones. The last section of this chapter lists some references that you can turn to if you want to explore further. But before concluding this section, a few remarks on some ideas and terms related to bifurcations.

6.5 Catastrophes and Tipping Points

There are two terms that are closely related to bifurcations: catastrophes and tipping points. In this section I'll briefly discuss both of these terms, comparing and contrasting them to bifurcations. A closely related phenomenon, phase transitions, is discussed in the next chapter in Section 7.8.

Catastrophe theory is an extrapolation of the theory of bifurcations. Developed in the late 1960s by René Thom, it received widespread attention when Thom published a book on catastrophe theory, *Structural Stability and Morphogenesis*, in 1972 (Thom, 1994). At a technical level, catastrophe theory sought to classify and explain jumps and discontinuities in mathematical models

---

3. Mathematically, one can locally transform a system near the equilibrium point into normal form via a Taylor expansion and a re-scaling of the variables. See Strogatz (2001, pp.48–51) for an example.

controlled by two or more parameters. But the aims of Thom were broader. He, and later Zeeman (1976; 1977), sought an anti-reductionist approach to modeling and a generic theory of forms and structures in natural and social systems. The idea was to study transitions and shifts in fairly abstract and geometric ways, as opposed to starting an analysis with equations obtained from first principles.

The 1970s and 80s saw an explosion of interest (and hype) around catastrophe theory. Several popular books appeared: Poston and Stewart (2012), Woodcock and Davis (1978), and Arnol'd (2003). The first two of these books were originally published in 1978; the book by Arnol'd first appeared (in Russian) in 1981. While catastrophe theory captured the imagination of many, both inside and out of math and science communities, it was met with strenuous objection and antagonism (Zahler and Sussmann, 1977; Kolata, 1977). What is the status of catastrophe theory today?

David Aubin writes bluntly that "Catastrophe theory is dead" (Aubin, 1998, p. 109), noting that there is almost no one who would describe her or himself as working on catastrophe theory, nor are there journals or conferences devoted to it. But Aubin goes on to argue that the influence of catastrophe theory lives on, largely as a matter of style or philosophy. René Thom, the principal founder of catastrophe theory, was interested not only in new mathematics but in "new ways to use mathematical tools and practice in order to make sense of the world (Aubin, 1998, p. 111)." This new way of understanding entails a focus on forms and structures themselves, without reference (at least initially) to the detailed equations of motion or the properties of the materials or entities under study.

This account is a rather simple sketch of what is a rich and layered episode in mathematics and science. For a short, semi-technical overview of catastrophe theory written during its heyday, see Zeeman (1976). Balanced accounts of the history and impact of catastrophe theory are Ekeland (1990), Aubin and Dahan Dalmedico (2002, Section 2.3), and Aubin (1998, Chapter 3). For discussions of the philosophy and epistemology underlying catastrophe theory, see Aubin (2001) and Aubin (2004).

We now turn our focus to *Tipping Points*, a term whose use has become quite common after the 2002 publication of Malcolm Gladwell's book by the same name 2002. A tipping point is a condition or threshold at which a change suddenly occurs—the system "tips" from one behavior to another. My sense is that "tipping point" is used mainly to describe transitions in social systems. That said, the notion of a tipping point is similar to the bifurcation: it is a sudden transition as some parameter or feature of a system changes. To my knowledge "tipping point" does not have a standard technical meaning in the way that bifurcation does. So you should be aware that tipping point is not a technical term and, in my experience, is not commonly used by those studying dynamical systems.[4]

By the way, it is interesting (to me, at least) to track the frequency of use of the terms "bifurcation", "catastrophe theory", and "tipping point". This can be done via a google ngram (Michel et al. (2011)), which lists the frequency of occurrence of words or phrases in a corpus of over a million books that google has digitized. The results are normalized to account for the fact that the number of books published yearly has increased with time. Such an ngram is shown in Fig. 6.7. The frequency of "catastrophe theory" rose from zero with the publication of Thom's book in 1972. The frequency of "bifurcation," being a word commonly used outside of mathematics, is fairly constant over time. The peak in the late 1940s is presumably due to frequent reference to the bifurcation of Europe.

---

4. Moreover, many of some of the claims put forth by Gladwell have been strongly (and in my view convincingly) critiqued. In particular, the idea that "influencers" or hubs play an essential role in the spread of ideas has met much criticism. See, e.g., Watts and Dodds (2007); Thompson (2008); Watts (2012).

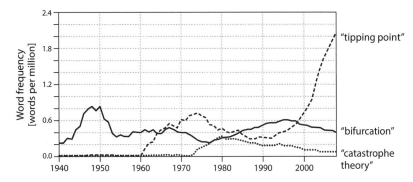

Figure 6.7. Google ngrams (Michel et al., 2011) for the phrases "bifurcation", "catastrophe theory" and "tipping point." http://books .google.com/ngrams.

As expected, the use of "tipping point" spikes upward around 2000. Gladwell's book was published in 2002, and he used the phrase in the title of a *New Yorker* article that appeared before the release of his book (Gladwell, 1996). The term "tipping point" was first used several decades earlier, to refer to the tendency for a neighborhood to "tip" from predominantly white to all black, once a certain fraction of homes in the neighborhood were owned by black families. This fraction was referred to as a "tipping point" (Griffith, 1961, note 67). A phenomenon closely related to bifurcations are *phase transitions*, which I discuss in Section 7.8 in the next chapter.

6.6 Hysteresis

A dynamical system with bifurcations can sometimes lead to a phenomenon known as hysteresis—a type of path-dependence or irreversibility. In this section I'll briefly work through an example of such a system and will offer a few thoughts on the importance and implications of hysteresis. This example follows the presentation in Strogatz (2001, pp. 59–62). This section is a bit more advanced and specialized than the rest of this chapter, and can be skipped or skimmed if the reader wishes.

Consider the differential equation:

$$\frac{dP}{dt} = rP + P^3 - P^5 . \tag{6.6}$$

Here P need not necessarily represent a population, but I will stick with the convention of this chapter and use P as the variable. This differential equation has one parameter, r. We will see how the solutions to this differential equation change as we vary r. As we did for the logistic differential equation in Fig. 6.2, we will plot the right-hand side of Eq. (6.6) for different r values and then draw the phase line. (In the logistic differential equation, Eq. (6.2), h was the parameter that we varied, while r and K were held constant.)

We'll start with $r = 0.01$. The right-hand side of the differential equation is graphed in the top plot of Fig. 6.8. We see that there are three equilibria: an unstable fixed point at $P = 0$ and two stable fixed points at $P \approx \pm 1$. At $r = -0.2$, shown in the middle of Fig. 6.8, the situation has changed. There are now five equilibria. The equilibrium at the origin, which formerly was unstable is now stable. There are two unstable equilibria at $P \approx \pm 0.6$, and at $P \approx \pm 0.85$ there are two stable equilibria. At $r = -0.28$, plotted in the bottom graph in Fig. 6.8, the situation has changed again. Now there is only one fixed point: a stable equilibrium at $P = 0$. Note that in Fig. 6.8 I have plotted stable equilibria with black circles and unstable equilibria with gray circles.

Note that all three phase lines in Fig. 6.8 are qualitatively different; they have different numbers of fixed points. Thus, a bifurcation must occur between $r = 0.01$ and $r = -0.2$ (moving from the top figure to the middle), and another bifurcation must occur between $r = -0.2$ to $r = -0.28$ (from the middle figure to the bottom). In Fig. 6.9 I have plotted the bifurcation diagram for Eq. (6.6). It is likely worth taking a moment to see how three phase lines in Fig. 6.8 can be found in the bifurcation diagram.

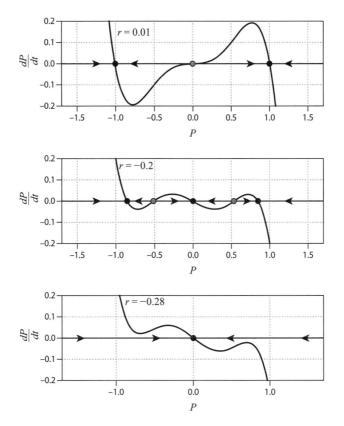

Figure 6.8. The right-hand side of Eq. (6.6) for three different values of r. The phase line for each dynamical system is drawn on the horizontal axis.

As in Fig. 6.5, stable equilibria are drawn in black and unstable equilibria in gray.

As anticipated, there are two bifurcations on this bifurcation diagram. There is one bifurcation at $r = 0$. The system shifts from five equilibria to three as r crosses 0 from the left to the right. There is another bifurcation at $r = -0.25$. If $r < -0.25$ there is only one stable equilibrium at $P = 0$ But if r is between -0.25 and 0, there are five equilibria. A population (or whatever) described

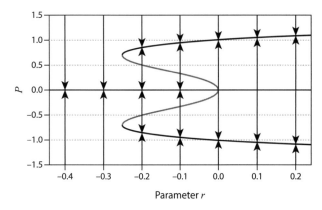

Figure 6.9. The bifurcation diagram for Eq. (6.6).

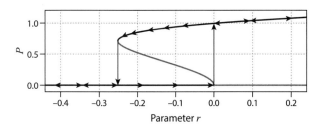

Figure 6.10. The bifurcation diagram for Eq. (6.6), for non-negative values of P. This diagram shows hysteresis, or path dependence. For parameter values between $r = -0.25$ and $r = 0$ there are two stable equilibria. In this region, which equilibrium the system is found in will depend on the path taken to arrive at that r value.

by this equation will undergo jumps, but will also exhibit a type of irreversibility.

To see this, we'll consider an example in which negative values of P are not possible—perhaps P is a population, and so negative populations are not meaningful. The bifurcation diagram for non-negative P is shown in Fig. 6.10. Let's suppose that initially the parameter r is 0.1. For this value there is one and only one equilibrium at a P that is slightly larger than 1.

Now suppose that we decrease the parameter r. The equilibrium value will decrease. If we decrease r a bit further, the equilibrium value for P decreases a bit further. This continues, until we get to $r = -0.25$. At this parameter value the equilibrium value for P disappears, and P plummets to zero. This is the same sort of bifurcation that we saw with the fish population in Fig. 6.5: an equilibrium value does not decrease gradually to zero, but jumps to zero at some critical parameter value.

Suppose that $r = -0.3$ and P is at zero. What happens if r is then increased? The system will remain at the equilibrium at $P = 0$, even if r gets larger than -0.25, the parameter value at which the crash occurred. The reason for this is that the equilibrium at $P = 0$ remains attracting as one crosses the critical value of $r = -0.25$. It is only when r is increased further, to $r = 0.0$, that the $P = 0$ equilibrium loses stability, at which point the system will jump back up to the stable equilibrium at $P = 1$.

The story I just told is illustrated in Fig. 6.10. If r is greater than zero, the system will be at the stable equilibrium near $P = 1$. If r is changed, the system moves back and forth along the upper branch of the bifurcation diagram. If r becomes less than zero, no sudden transition occurs. The value of P continues to decrease smoothly. A sudden transition will occur, however, the moment that r dips below 0.25. At this point P jumps to zero. The story is different if one moves across the diagram in the opposite direction, from left to right. As r is increased, there is no transition at $r = -0.25$; the value of P remains at zero. But a transition does occur later on; P jumps from 0 to 1 at $r = 0$.

The behavior of this system shows *path dependence*. In between $r = -0.25$ and $r = 0$, the value of P depends on the path the system took to get there. This phenomenon of path dependence is also known as *hysteresis*. It occurs often in systems with multiple stable equilibria, as is the case in Fig. 6.10 when r is between -0.25 and 0.0. For such a system, knowing the value of the parameter is not enough to determine its behavior. For example, if I told you at $r = -0.1$, you would not be able to figure out the value of P. You would need to know the path the system took—how did it come to have this r value? Did it move along the upper or lower equilibrium? History matters. In contrast, if $r = 0.1$, you can be certain that the population P will be approximately one, regardless of what r values it had previously.

Systems with hysteresis possess a simple type of memory. In the region between $r = -0.25$ and 0 in Fig 6.10, the system "remembers" where it has been. This is another surprising feature of simple differential equations. The differential equation we're working with in this section, Eq. (6.6), is autonomous—dP/dt, the rate of change of P, depends only on P, not on the time t. So the equation appears to be memoryless; it does the same thing regardless of the value of time t. However, as we have seen, if one varies the parameter r, a type of memory or path dependence occurs. It is perhaps a long way from Eq. (6.6) to any realistic model. But I think this example is useful nevertheless. It provides intuition about how hysteresis or path dependence can occur when a system has multiple equilibria.

6.7 Further Reading

Bifurcations are a standard topic in dynamical systems and are also treated in some differential equations courses. Strogatz (2001, Chapters 3 and 8) and Kaper and Engler (2013, Chapter 5) both give particularly clear treatments of bifurcations. Kaper and Engler (2013) also contains several examples of bifurcations arising in models of the earth's climate. Appendix B of Hilborn (2002) is a short introduction to bifurcation theory, including a mention of the center manifold theory. More advanced and formal treatments can be found in Crawford (1991) and Guckenheimer and Holmes (2013).

7

UNIVERSALITY IN CHAOS

In the previous chapter we looked at bifurcation diagrams for differential equations. In this chapter we will return to iterated functions and examine their bifurcation diagrams. We will see that such diagrams are remarkably complex—much more so than the bifurcation diagrams for differential equations that were the topic of the previous chapter. Looking at the regularities and structures in these bifurcation diagrams will lead us to the notion of universality, in my opinion one of the most amazing results to emerge from the study of dynamical systems.

7.1 Logistic Equation Bifurcation Diagram

Our main example in this chapter will be the bifurcation diagram for the logistic equation:

$$f(x) = rx(1-x). \tag{7.1}$$

This equation was introduced in Section 4.1 as a very simple model for a population that has some limit to its growth. We saw in Sections 4.2 and 4.3 that different values of the growth parameter r led to time series with a range of behaviors, both periodic and aperiodic. The main results of our investigations were summarized in Fig. 4.5. For convenience, I've reproduced these plots in Fig. 7.1. In the previous chapter we formed a bifurcation diagram

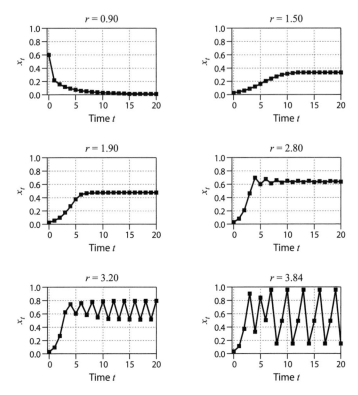

Figure 7.1. Time series for the logistic equation for six different r values. This figure is identical to Fig. 4.5.

by "gluing" together a bunch of phase lines for a differential equation. Bifurcation diagrams for iterated functions are formed via a similar strategy.

For iterated functions, however, one summarizes the behavior of the system with a construction known as a *final-state diagram* and not a phase line, as was the case for differential equations. The idea behind a final-state diagram is, as the name suggests, to record the final state or states of the dynamical system. For example, if $r = 0.9$, we see in the upper left plot of Fig. 7.1 that the orbit approaches zero. The final state in this case is just the point

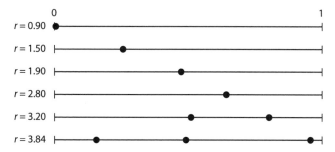

Figure 7.2. Final-state diagrams for the logistic equation for several different values of the parameter r. Orbits for these r values are shown in Fig. 7.1.

$x = 0$. If we iterated the function for a while, the orbit will end up stationary at $x = 0$. This is illustrated in the top diagram of Fig. 7.2.

If $r = 1.50$ the situation is slightly more interesting. We see in Fig. 7.1 that the orbit is pulled toward an attracting fixed point at $x = 0.33$. The final state in this case is just the single point $x = 0.33$. This is shown in the second diagram in Fig. 7.2. In general, if the orbit is pulled toward an attracting fixed point, then the final-state diagram is a just a single point—namely, the fixed point. We can see this for $r = 1.90$ and $r = 2.80$ in Fig. 7.2.

What if the orbit is pulled toward a periodic cycle instead of a fixed point? This is the case for $r = 3.20$, as shown in the lower left of Fig. 7.1. Here the orbit oscillates between $x \approx 0.52$ and $x \approx 0.8$. In this case there are two final states, not one, as shown in the final-state diagram in Fig. 7.2. Similarly, when $r = 3.84$, shown in the bottom right of Fig. 7.1, the orbit is pulled toward a cycle of period three, and there are three values on the final-state diagram.

The final-state diagrams for periodic orbits are fairly straightforward—the diagram consists of a number of dots, one for each point on the periodic cycle. But how can we represent chaotic behavior on a final-state diagram? In Fig. 7.3 I have shown the orbit for the logistic equation with $r = 4.0$, a parameter value

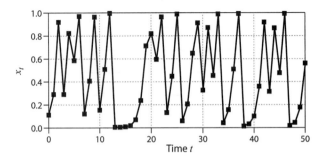

Figure 7.3. An orbit for the logistic equation with $r = 4.0$. For this parameter orbits are aperiodic; they do not repeat.

Figure 7.4. The final-state diagram for the logistic equation with $r = 4.0$. The orbit is shown in Fig. 7.3.

that we have seen yields chaos. How can we represent this behavior? Since the orbit is chaotic, it is aperiodic; it doesn't become periodic.

The answer to this puzzle is to simply plot a few hundred points of the itinerary on the final-state diagram. This is a reasonable representation of the final "state" of the system. After all, the orbit doesn't settle down; it keeps jumping around, never returning to exactly the same place. So its final-state diagram should reflect this perpetual wandering. Such a final state diagram is shown in Fig. 7.4. The diagram includes around 40 points. If I plotted more points, the dots on the line would get denser and denser, until it appeared as a solid smear of black. The points never land exactly on top of each other, since the orbit never exactly repeats. But the orbit bounces all over the interval between 0 and 1. Thus, since the dots on the diagram are of finite size, eventually the dots would fill up the entire diagram.

The general algorithm for making a final-state diagram is as follows. Choose the parameter value r and a seed x_0 for the orbit.

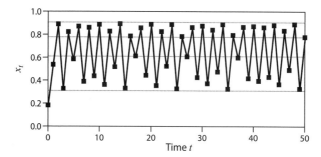

Figure 7.5. An orbit for the logistic equation with $r = 3.61$. For this parameter orbits are aperiodic; they do not repeat. However, the orbit does not take all values between 0 and 1, as was the case for $r = 4.0$.

Then iterate the function for, say, 200 time steps. This ensures that the system has moved beyond any transient behavior and is in its final state, be it a periodic or an aperiodic one. After this, iterate for 100 more timesteps and plot those 100 iterates as dots on the interval [0, 1]. This recipe will yield a final-state diagram.[1]

Before we use final-state diagrams to make a bifurcation diagram for the logistic equation, there is one more subtlety to consider. In Fig. 7.5 I have plotted the time series for the logistic equation with $r = 3.61$. For this parameter value the orbits are chaotic. If you look closely you will see that the orbit in the figure is aperiodic. (Iterating for more timesteps will confirm this.) However, the orbits do not wander all over the unit interval, as they did for the $r = 4.0$ case. Instead, in the long run the orbit in Fig. 7.5 is confined to two bands. There is a large lower band, from

1. This recipe may need some adjustment, however. There could be long transients—it might take the system more than 200 iterates to reach its final behavior. Also, depending on the resolution one desires, one might need to plot many more than 100 iterates. This algorithm is not efficient for making final-state diagrams for periodic orbits. For example, if the orbit moves to a fixed point, then the 100 iterates plotted would all be essentially the same, and so plotting all 100 of them would be wasteful and slow. See footnote 2 for more discussion of how to make bifurcation diagrams more efficiently.

Figure 7.6. The final-state diagram for the logistic equation with $r = 3.61$. The orbit is shown in Fig. 7.5.

$x \approx 0.31$ to $x \approx 0.61$, and a smaller upper band, from $x \approx 0.78$ to $x \approx 0.89$. To help make this clearer, on Fig. 7.5 I have drawn horizontal lines at $x = 0.31, 0.61, 0.78$, and 0.91. One sees that except for the initial value, all iterates fall between the pairs of gray lines.

This is reflected in the final-state diagram for the logistic equation with $r = 3.61$, shown in Fig. 7.6. We see that the orbits fall within two bands. If I plotted more orbits, the two bands would be completely filled in. The orbits are aperiodic—they do not repeat. They keep bouncing around the two bands. Note, by the way, that there is some regularity to the orbit. The iterates alternate between bands. If an iterate is somewhere in the lower band, the next iterate will always be in the upper band, and vice versa. The orbits for this r value are aperiodic and have sensitive dependence on initial conditions, but the orbit is not completely patternless.

We are now ready to make the bifurcation diagram. This will allow us to visualize how the system's long-term behavior changes as the parameter r changes. Already we have seen a number of different behaviors, periodic and aperiodic. What else can the logistic equation do, and how does its behavior change as r is varied? We'll begin constructing the bifurcation diagram by using the final-state diagrams for the eight r values we have considered so far: $0.90, 1.50, 1.90, 2.80, 3.20, 3.61, 3.84$, and 4.00. The result of doing this is shown in Fig. 7.7. I have taken the final-state diagrams and rotated them ninety degrees counter-clockwise. I have also drawn an r-axis horizontally and glued the final-state diagrams at the appropriate location. For example, the final-state diagram for $r = 1.90$ (shown in Fig. 7.2, third from the top), is placed at

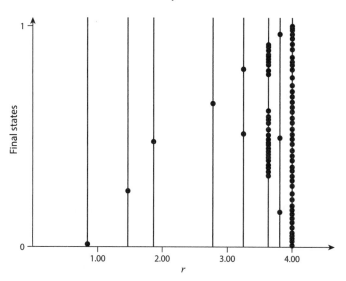

Figure 7.7. The first steps in the construction of the bifurcation diagram for the logistic equation. Plotted on the horizontal r-axis are the final-state diagrams from Figs. 7.2, 7.4, and 7.6.

$r = 1.9$ on the r-axis of Fig. 7.7, just to the left of the $r = 2.00$ marker. You might want to take a moment and see where the other final-state diagrams from Figs. 7.2, 7.4, and 7.6 have been placed in Fig. 7.7.

It is hard to get a full picture of the logistic equation's behaviors from the bifurcation diagram of Fig. 7.7. Since what we're after is a picture of all the behaviors the equation exhibits and how those behaviors change as r is varied, we will need to choose many more r-values, determine the final-state diagram for each, and add them to the bifurcation diagram. At this point we'll turn to a computer to make the bifurcation diagram for us. The result of doing so is shown in Fig. 7.8.

Here's how I made this figure. I asked my computer to do the following for 1000 r values between 0 and 4.0: iterate the logistic equation for 200 time steps; then iterate for 200 more and

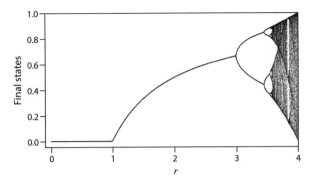

Figure 7.8. The bifurcation diagram for the logistic equation.

plot those 200 points on the bifurcation diagram. This has the effect of making a finite-state diagram for 1000 different r values. Plotting them all together yields the bifurcation diagram shown in Fig. 7.8.[2]

2. Actually, while the description I just gave for making the bifurcation diagram will work, I did something a bit more complicated so the program ran faster. If an orbit is periodic, there is no need to plot many iterates when making a final-state diagram. A periodic orbit repeats, and so fairly soon the points on the graph fall on top of each other, providing no additional information but still taking up computer time and making the plot an unnecessarily large file. The program I wrote detected if the orbit was periodic and only plotted 10 iterates in this case. Otherwise it plotted 200 iterates.

To detect whether or not the orbit was periodic, I calculated the Lyapunov exponent. For an iterated map f, it is a standard result that the Lyapunov exponent λ can be calculated from an itinerary $x_0, x_1, \ldots x_n$ via the following formula:

$$\lambda = \lim_{N \to \infty} \frac{1}{N} \sum_{i=0}^{N} \ln(f'(x_i)). \tag{7.2}$$

(See, e.g., Peitgen et al. (1992, Section 10.1) or Strogatz (2001, Section 10.5).) Equation 7.2 thus gives an efficient way to estimate the Lyapunov exponent; one just iterates for a little while, evaluating the derivative of f at each iterate, and putting the log of the result into the sum. As mentioned in Section 4.7, for a chaotic system, the Lyapunov exponent λ is positive, whereas it is negative for a periodic system. Hence, calculating λ using Eq. (7.2) and then examining its sign gives us a quick way to tell if the orbit of an iterated map is periodic or chaotic.

Before proceeding to explore the bifurcation diagram, a few remarks on terminology. The plot shown in Fig. 7.8 is very commonly referred to as a bifurcation diagram. Note, however, that Fig. 7.8 is very different from the bifurcation diagrams for differential equations explored in the previous chapter. Those bifurcation diagrams showed both stable and unstable fixed points, whereas Fig. 7.8 only includes stable behavior. The bifurcation diagrams of this chapter are often referred to as *orbit diagrams* (e.g., Strogatz (2001)), since what they show is the long-term behavior of the orbit for each r value. They are also sometimes calls *final-state diagrams* (e.g. Peitgen et al. (1992)). I will continue to call figures like Fig. 7.8 a bifurcation diagram. This is slightly sloppy terminology, but it is also quite standard and I think is unlikely to cause confusion. For an engaging discussion of the "real" bifurcation diagram with unstable fixed points included, as opposed to the "impostor" orbit diagram, see Ross and Sorensen (2000) and Ross et al. (2009).

7.2 Exploring the Bifurcation Diagram

As discussed in Chapter 6, a bifurcation diagram serves as a dictionary of possible behaviors for a dynamical system. If you want to know the behavior for a particular parameter value, locate that parameter on the horizontal axis of the bifurcation diagram. A vertical slice directly above that point is the final-state diagram for that parameter value. For example, in the bifurcation diagram of Fig. 7.8 we see that if $r = 2.0$, the final-state diagram is a single point at a bit less than $x = 0.6$. This indicates that the logistic equation with $r = 2.0$ has a single, attracting fixed point at $x \approx 0.6$.

The bifurcation diagram of Fig. 7.8 reveals a range of behaviors. When r is between 0 and 1, there is a single attracting fixed point at 0. This makes sense; if the growth rate r is less than one, the population will always die off. For $r = 1$ to $r = 3$, there is a non-zero fixed point; the population reaches a steady value. This steady value increases as the parameter r increases. Then when r is greater than 3.0, the line on the bifurcation diagram splits in two. This indicates that we now have period-two behavior; in the long run orbits are pulled toward an attracting cycle of period two. As r increases further still, the two lines in the bifurcation diagram split again, indicating period-four behavior.

The change in behavior that occurs at $r = 3.0$ is known as a *period-doubling bifurcation*. Recall that a bifurcation is a sudden, qualitative change in a dynamical system's behavior as a parameter is varied continuously. The qualitative change seen here is that the period of the attractor doubles. If r is a bit less than 3.0, the logistic equation has an attracting fixed point. If r is a bit larger than 3.0, then the logistic equation has a periodic attractor of period two. There is another period-doubling bifurcation at $r \approx 3.449$. If r is a bit less than this value, the attractor has period two; if r is a bit larger than this value, the attractor has period four. Period-doubling bifurcations are also known as *pitchfork bifurcations*.

I'll have more to say about period-doubling bifurcations in the next few sections. For now, let's keep exploring the bifurcation diagram. In Fig. 7.8 we see period-two behavior for r between 3.0 and around 3.5, and after that we see period-four behavior. Then what? It is hard to see on this figure. Let's take a closer look. Figure 7.9 shows the bifurcation diagram for the logistic equation for r ranging from 3.0 to 4.0. We see that the bifurcation diagram has a complicated structure. The equation is chaotic for many parameter values—all those above for which there is a solid or almost-solid vertical line. (Recall that for a chaotic dynamical system the orbit is not periodic, and so the final-state diagram appears as one or more solid regions, as in Fig. 7.4 and 7.6.) Amidst these chaotic regions there are also many regions where the orbits are again periodic. These are known as *periodic windows* and appear as narrow vertical gaps in the darker, chaotic regions. For example there is a window of period three around $r = 3.83$.

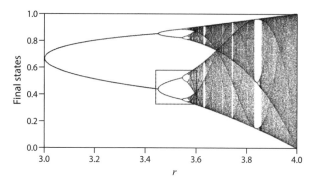

Figure 7.9. The bifurcation diagram for the logistic equation, $r = 3.0$ to 4.0.

To get a better sense of what is going on in the bifurcation diagram, let's take a closer look and zoom in. Figure 7.10 shows the bifurcation diagram for smaller and smaller ranges of r and x. The top plot in this figure shows the region in Fig. 7.9 that is marked with a rectangle. The middle plot of Fig. 7.10 shows the small boxed region in the top plot. And the bottom plot shows the region of the bifurcation diagram that is boxed in the middle plot. If the entire bifurcation diagram of Fig. 7.9 was expanded as much as in the bottom plot of Fig. 7.10, it would be around 100 meters wide. The self-similar, fractal structure of the bifurcation diagram is apparent; we see similar shapes appearing at many different scales.

Figure 7.10 reveals a remarkable structure. We see that the logistic equation's orbits shift from chaotic to periodic and back again as r is varied. This is another sense in which a chaotic system can exhibit sensitive dependence. The dynamical system depends sensitively on the value of the parameter; small differences in the value of r can lead to qualitatively different behavior. In fact, periodic windows in the bifurcation diagram are dense. What this means is that in *any* interval of chaotic r-values there is an interval of r values for which the orbits are periodic (Peitgen et al., 1992, p. 638).

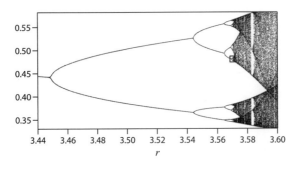

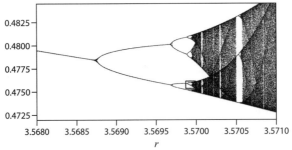

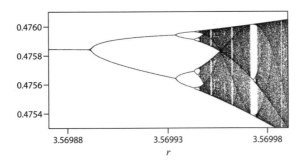

Figure 7.10. Successive zoom-ins of the bifurcation diagram for the logistic equation.

So whenever one zooms in on a region that appears solidly chaotic, one will see small periodic regions. Another way to say this is that there are infinitely many periodic windows between any two periodic windows.

And yet, there is quite clearly regularity and structure in the bifurcation diagram. One can see repeating patterns as one zooms in. I'll examine one of these regularities—the sideways U's indicating successive period-doubling bifurcations—in the next section. Before doing so, a few remarks on emergence.

7.3 Some Words about Emergence

The handful of logistic equation bifurcation diagrams that I've plotted in this chapter are just the tip of the iceberg. I'd encourage you to explore further using one of the many programs online that will plot bifurcation diagrams for you. Searching will bring up many such programs[3], or you can write your own. It is fun to choose what appears to be a solidly chaotic region of the bifurcation, zoom in, and discover periodic windows. It is also fun to zoom in on the right-hand side and admire the veils and ribbons that decorate the bifurcation diagram.

It's worth reminding ourselves that all the structures in the bifurcation diagrams—all the plots so far in this chapter and all of those you might make exploring on your own—come from iterating a simple equation:

$$f(x) = rx(1 - x). \tag{7.3}$$

Moreover, the code to generate the bifurcation diagram is short and simple. One basically iterates the equation, discards the first

---

3. One of my favorite such programs can be found at: https://s3.amazonaws .com/complexityexplorer/DynamicsAndChaos/Programs/bifurcation.html.

hundred or so iterates and keeps the next hundred. Do this for a range of r values, and plot. That's it.

And yet there is a powerful sense in which writing down the logistic equation or even specifying the bifurcation-diagram-generating algorithm doesn't really explain the bifurcation diagram. It feels like a long way from Eq. (7.3) to the plots in Fig. 7.10. This is arguably an example of an *emergent phenomenon*, which is informally defined as a pattern or behavior that is not directly contained in the description of the pattern's constituents and their interactions.

In my opinion emergence is a difficult notion to nail down rigorously and is sometimes discussed with unneeded mysticism or fanfare. But intuitively, emergence is palpable. I think of it as the feeling of not knowing where something came from, even if you know where it came from. By this I mean the following. On the one hand, I know exactly where the bifurcation diagrams in this chapter come from. After all, I wrote the program that generated the plots. But when looking at the figures, and especially when zooming in repeatedly, one gets the sense that it doesn't quite seem right to say that I made the bifurcation diagram. Where *does* the pattern come from?

Perhaps it is best to think of the pattern as being generated by the act of running the code, not writing it. In this sense, making a bifurcation diagram is very different from, say, building a bookcase. If I were to build a bookcase, I would first draw up a sketch of what I wanted the bookcase to look like. Then I would build something that looks (hopefully) like the figure I just sketched. In contrast, in building a bifurcation diagram, I definitely don't produce a sketch or representation of the diagram. Instead, I specify the algorithm that will make the bifurcation diagram, and let it run. So the pattern emerges from the dynamics but is not directly present in the dynamical rule itself. I'll return to the notion of emergence in Section 10.3.

7.4 The Period-Doubling Route to Chaos

We now turn our attention to one of the more striking regularities in the bifurcation diagram: the successive period-doubling bifurcations that are responsible for the sideways U-shapes or pitchforks in the bifurcation diagram. Doing so will lead us to the phenomenon of universality. We shall see that there are certain features of chaotic transitions that are the same for almost all dynamical systems, and which can be observed in physical systems, as well.

In Fig. 7.10 we see period-doubling occurring throughout the bifurcation diagram at different scales. As the parameter r is increased, the period doubles, then doubles again, and so on. The distance between period doublings gets smaller and smaller as r gets larger. Eventually, the period doublings "*accumulate*" and the dynamical system transitions to chaotic behavior. These successive period doublings are often referred to as a *period-doubling cascade* or the *period-doubling route to chaos*. The parameter value at which the function shifts from periodic to chaotic is known as the *accumulation point*.

There is a regularity to the period-doubling route to chaos that forms the basis of one of the most remarkable results from the study of dynamical systems. This regularity can be captured as follows. Figure 7.11 shows a sketch of three period-doubling bifurcations: from period one to period two, period two to period four, and period four to period eight. The period-eight behavior would then split into period-sixteen behavior on the far right of the diagram, although this transition is not shown. The approximate parameter values at which these bifurcations occur are listed in Table 7.1.

We can use the data in the table to determine the widths of each periodic region in Fig. 7.11. For example, the width of the period-two region, denoted Δ_1, is given by:

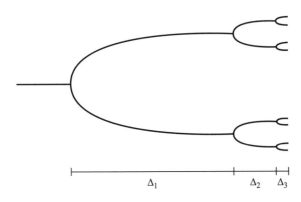

Figure 7.11. A sketch of several period-doubling bifurcations.

$$\Delta_1 = 3.44949 - 3.00000 = 0.44949 . \qquad (7.4)$$

A similar calculation yields $\Delta_2 \approx 0.09460$ and $\Delta_3 \approx 0.02002$. The region that has period 2 is larger than the period-four region: Δ_1 is larger than Δ_2. How many times larger? Well,

$$\frac{\Delta_1}{\Delta_2} \approx \frac{0.44949}{0.09460} \approx 4.7515 . \qquad (7.5)$$

So the period-two region on the bifurcation diagram is about 4.75 times larger than the period-four region. I'll denote this ratio by δ_1. That is, $\delta_1 = \Delta_1/\Delta_2$.

We also see on Fig. 7.11 that the period-four region is bigger than the period-eight region. How many times larger? Denoting this ratio by δ_2:

$$\delta_2 = \frac{\Delta_2}{\Delta_3} \approx \frac{0.09460}{0.02002} \approx 4.7253 . \qquad (7.6)$$

Hmmm. The two ratios are about the same. If we continue exploring further period doublings in the bifurcation diagram, we will continue to see approximately the same ratio. The width of each period-doubling region is around 4.7 times larger than the next.

Stating this more precisely, let δ_n be the ratio of the nth and the $(n+1)^{\text{th}}$ periodic regions: $\delta_n = \Delta_n/\Delta_{n+1}$. Then it turns out that

Transition	r
$1 \rightarrow 2$	3.00000
$2 \rightarrow 4$	3.44949
$4 \rightarrow 8$	3.54409
$8 \rightarrow 16$	3.56441

Table 7.1 The approximate r values at which the first several period-doubling bifurcations occur for the logistic equation.

$$\lim_{n \to \infty} \delta_n \approx 4.669201609 . \tag{7.7}$$

In words, as one goes farther and farther along the period-doubling route to chaos, the ratio of the widths of successive periodic regions approaches 4.669201609. The number 4.669201609, which I'll round to 4.669 from now on, is known as Feigenbaum's constant. In the late 1970s, Pierre Coullet and Charles Tresser (1978; 1980) and Mitchell Feigenbaum (1978) independently realized that this number is relevant to much more than the logistic equation, as we shall see.

7.5 Universality in Maps

So far in this chapter we have looked at the bifurcation diagram for the logistic equation in quite a bit of detail, but what about other iterated functions. Are their bifurcation diagrams similar to that of the logistic equation, or are all bifurcation diagrams completely different? Let's look at the bifurcation diagram for a function other than the logistic equation. For example,

$$f(x) = axe^{-x} . \tag{7.8}$$

I'll refer to this as the exponential function, even though it is not a pure exponential e^{-x}. The variable a is a parameter, just as r is a parameter for the logistic equation.

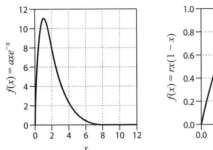
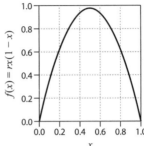

Figure 7.12. Plots of the logistic and exponential functions. The parameter values are $a = 30$ and $r = 3.9$.

A plot of this function is shown in the left half of Fig. 7.12. The right plot is the familiar logistic equation, $f(x) = rx(1 - x)$. Note that both functions have a single maximum and both have bounded orbits. For the exponential function we can see on Fig. 7.12 that since the maximum value of the function is around 11, any initial x value between 0 and 11 will remain between 0 and 11. (The exact value of the maximum is $30e^{-1} \approx 11.036$.)

What does the bifurcation diagram for the exponential function look like? The answer to this question is found in Fig. 7.13, which shows a bifurcation diagram I made using the same methods used to produce the logistic equation bifurcation diagrams earlier in this chapter. The exponential bifurcation diagram looks a lot like the bifurcation for the logistic equation, shown in Fig. 7.8. This is curious. The two bifurcation diagrams definitely aren't identical, but both show sideways U's indicating that the function is undergoing period doubling—the structure sketched in Fig. 7.11. What if we calculated the δ's, the factors by which the width of one periodic region is larger than the next one. Doing so, I find the following:

$$\delta_1 \approx 2.99\,, \tag{7.9}$$

$$\delta_2 \approx 4.21\,, \tag{7.10}$$

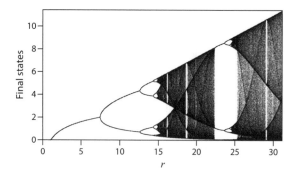

Figure 7.13. The bifurcation diagram for the exponential function, Eq. (7.8).

$$\delta_3 \approx 4.57, \tag{7.11}$$

$$\delta_4 \approx 4.64. \tag{7.12}$$

This is curiouser still. It looks like δ for the exponential function is approaching the same number that it did for the logistic equation.[4] It seems that these two different functions have something in common.

It turns out that the number that the δ's approach, 4.669, is *universal*. We could repeat this experiment with more and more functions and we would keep finding δ's that approach 4.669. Let me state this result a bit more carefully, and then I'll talk about what this result might mean. Given a sequence of period doublings, define δ_n as the width of the nth periodic region divided by the width of the $(n+1)^{\text{th}}$ region. Then $\lim_{n\to\infty} \delta_n = 4.669$

4. By the way, it can be difficult to pin down the precise numerical values for the parameter values at which bifurcations occur. At the bifurcation point the attractor shifts from one period to the other. In the vicinity of the bifurcation, the attractors are very weakly attracting, so it takes a long time for the orbit to get pulled into them, and so it is not easy to see what the period of the attractor is. This problem becomes more and more pronounced for longer periods. For example, while it is not too hard to determine the parameter value at which behavior shifts from period two to period four, it is very challenging to pin down the parameter at which the behavior shifts from period 32 to period 64.

for *any* iterated function with a quadratic maximum that exhibits a period-doubling route to chaos.[5] A quadratic maximum is a maximum that can be locally approximated by a parabola. More formally, it is a maximum that has a non-zero second derivative.

What the phenomenon of universality means is that certain features of the period-doubling route are the same for all functions. Almost all functions that undergo the period-doubling transition to chaos do so in a similar way. So there are similarities among a broad class of different dynamical systems. The logistic equation is just about the simplest non-linear equation imaginable; it's an upside-down parabola. Nevertheless, aspects of its period-doubling route to chaos have the same geometric characteristics as the period-doubling route to chaos for more complicated equations: not just the exponential function but *any* function with a single quadratic maximum. So this simple model, the logistic equation, captures properties of more complicated models. Moreover, this is a quantitative result. It's not just that the bifurcation diagrams look kinda similar; they share quantitative geometric properties such as the ratio δ. (The ratio of the heights of successive periodic regions is also universal, although I won't discuss this here.) How is universality possible? I'll discuss this some in Section 7.7. But first, let's take a look at some physical dynamical systems—real experiments—that exhibit period-doubling.

---

5. There is a bit more mathematical fine print. In addition to the requirement that f has a single quadratic maximum, the function must be smooth, its orbits must be bounded, and the function must have a negative Schwarzian derivative $S[f]$, where

$$S[f] = \frac{f'''}{f'} - \frac{3}{2}\left(\frac{f''}{f'}\right)^2 . \tag{7.13}$$

For a discussion of some of these mathematical details, see Schuster and Just (2006). In practice, my sense is that the Schwarzian derivative condition is almost always met. The upshot is that a generic or typical function with a single quadratic maximum will almost surely have the universal value for δ.

7.6 Universality in Physics

Period-doubling is not just a mathematical phenomenon; the period-doubling route to chaos has been observed in physical systems as well. I'll briefly discuss an example. Chaotic behavior can be observed in electrical circuits. One such circuit consists of a resistor, inductor, and a diode in series (Linsay, 1981). The capacitance of the diode introduces a time delay into the circuit, since it does not charge and discharge instantaneously. This capacitance depends nonlinearly on the voltage across the diode, and the circuit is driven by a sinusoidal voltage. When the amplitude of the driving voltage is low, the voltage across the diode has the same frequency as the driving frequency. As the voltage is increased, the circuit undergoes a bifurcation and the frequency of the voltage across the diode doubles. If the voltage is increased further still, the frequency across the diode doubles again, and so on. For a large enough driving frequency, the system becomes chaotic.

One can determine the voltages at which these bifurcations occur, just as one can determine the values of the parameter r or a at which bifurcations occur for the one-dimensional functions studied in this chapter. Linsay (1981) determined these voltages and then calculated values for δ. He found that $\delta_1 = 4.4 \pm 0.1$ and $\delta_2 = 4.5 \pm 0.6$, in good agreement with the theoretical value of 4.669. (The theory says that δ_n only equals 4.669 in the large-n limit. In practice, however, δ_n is often quite close to Feigenbaum's value of 4.669 for small n.)

There are many other physical systems that exhibit a period-doubling route to chaos as a parameter is varied. Examples include several other circuits, laser and acoustic feedback systems, oscillating chemical reactions, and oscillations in convection rolls of fluids. In all cases the δ's that are calculated are consistent with Feigenbaum's value of 4.669. A good summary of this experimental work is the edited volume by Cvitanović (1989).

This is a remarkable result. The physical systems that undergo period-doubling and are described by the universal value of 4.669 are quite different from one another: electrical circuits, convection rolls in fluids, oscillating chemical reactions, and so on. It is not at all obvious what these systems have in common. But they must have *something* in common, since their transitions to chaos are all captured by the universal value of δ. How can this be?

Moreover, what in the world do these physical systems have in common with one-dimensional iterated functions? I introduced the logistic equation in Section 4.1 as a deliberately almost-too-simple equation that models a population that has some limit to its growth. We noticed in the bifurcation diagram for this equation a geometric regularity in the period-doubling route to chaos; the ratio of the widths of the periodic regions converges to 4.669. And then we found this same number in multi-dimensional physical systems whose relation to one-dimensional iterated functions is not clear.

At this point there are two mysteries that need explaining.

1. In the previous section we saw that the period-doubling route to chaos is universal. All one-dimensional iterated functions with a single quadratic maximum are described by the ratio $\delta \approx 4.669$. Why?
2. What do one-dimensional iterated functions have to do with higher-dimensional systems such as electric circuits or turbulent fluids?

There are well understood answers to both of these questions. Let me begin with a few words about the second question. Geometrically, any system that exhibits chaotic behavior must involve both stretching and folding. One-dimensional functions with a single maximum also stretch and fold. This one-dimensional stretching and folding captures the same basic geometric action that occurs in some higher-dimensional chaotic systems. We'll return to this in Sections 9.4 and 9.5. We now turn our attention to the first question: why is the period-doubling transition to chaos universal for one-dimensional iterated functions?

7.7 Renormalization

There is a well-developed mathematical framework that helps one see why certain features of the period-doubling route to chaos in one-dimensional iterated functions are the same. Additionally, in this framework one can calculate values such as δ. This approach is known as *renormalization*. In what follows I'll give an sketch of renormalization and how it explains universality. Section 7.7.1 is a non-technical sketch of renormalization. In Section 7.7.2, I look at applying renormalization to period doubling. This section is fairly abstract and can certainly be skimmed or skipped. Section 7.7.3 is a quick summary of renormalization.

7.7.1 A Very Simple Example

Renormalization concerns how properties of a system change when the scale of the system is changed. I'll illustrate this first with a simple example. Consider a curve. If you zoom in on a portion of a curve, the curve will look straighter. And if you zoom in again, it will look straighter still. The first step of this process is illustrated in the top part of Fig. 7.14. If the shape we're zooming in on happens to be a straight line, then its appearance is unchanged by zooming in, as shown in the bottom part of the figure.

Let's introduce some terminology. The process of changing scale is known as *renormalization*. A particular zoom-in would be referred to as a renormalization operation. The framework of renormalization is sometimes called the *renormalization group*, since the collection of renormalization operations has the property of a mathematical structure known as a group[6]. If a shape

---

6. Strictly speaking, the renormalization group is not a always group. The zooming procedure described here is invertible, but many renormalization operations in other contexts are not invertible. So in general, the set of all possible

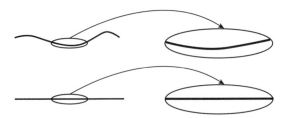

Figure 7.14. An illustration of a renormalization operation that consists of zooming in—magnifying the portion of a shape by a factor of three. In the top part of the figure, we see that zooming in on a curve makes it appear straighter. In the bottom figure, the curve is already straight, so zooming in on it does not change the shape.

is unchanged by the renormalization operation, then we call that shape a *fixed point*. The straight line, shown in the bottom of Fig. 7.14, is a fixed point of the renormalization operation in this example. One can view renormalization as a dynamical system and ask what happens to different initial conditions when successively renormalized. Are there any fixed points? Are they attracting or repelling?

As noted, curves become straighter when renormalized, as illustrated in the top of Fig. 7.14. As one zooms in more and more, the curve bears a closer and closer resemblance to a straight line. Thus, the straight line is an attracting fixed point. Curves, when renormalized, approach straight lines. The line might not be horizontal, but it will be straight. In the language of renormalization, one would say the straight line is a *universal* curve. It's universal in the sense that all other curves approach this universal curve—that is, the straight line—when renormalized. If an object is a fixed point of a renormalization operation, then this means that that object does not change appearance when the scale is changed. When this is the case, one says that the object is scale free.

---

renormalization operations has all of the properties of a group except it is not always invertible. A set with these features is called a *semi-group*.

Figure 7.15. Two curves that will not approach a straight line when zoomed in.

But there's a bit more to the story than this. There are some curves that can't be turned into straight lines. Two such curves are shown in Fig. 7.15. The curve on the left is discontinuous. If we zoom in near the discontinuity, the discontinuity remains. It will not look like a straight line under continuous zooms. The curve on the right is continuous but it is "pointy"—it has a non-continuous slope at the sharp peak. This pointiness persists as one zooms in. Again, the curve will not approach a straight line.

To put it all together, a renormalization operation entails changing the scale of view or analysis of the system under study. In this example the renormalization operation consists of magnification—zooming in. The operator acts on curves. A straight line is an attracting fixed point of the renormalization operator. All continuous and smooth curves will approach a straight line when repeatedly magnified.[7]

7.7.2 Renormalization Applied to Period Doubling Transition

We now need to talk about how to apply renormalization to the bifurcation diagram of the logistic equation. First, let's think about why we might want to apply renormalization in the first place. The process of renormalization is useful when there is some sort of scale-free behavior. The pitchforks in the bifurcation diagram are scale free. If you're in a periodic region on the bifurcation diagram, the next one in the doubling sequence will be 1/4.669 times

---

7. As it has perhaps occurred to you, this is the central idea behind differential calculus. The graph of a function $f(x)$ is a curve. If one zooms in on this curve around some point x, the curve will look more and more like a straight line. The slope of the straight line is the derivative of $f(x)$ at x.

smaller. As we zoom in and out of the bifurcation diagram we see this same pattern repeated across scales: pitchforks on which are 1/4.669 times smaller pitchforks, on which are 1/4.669 smaller pitchforks, and so on.

Since we observe scale-free behavior, we seek to apply renormalization. But what is there to renormalize? Well, since we see scale-free behavior in the bifurcation diagram, presumably there is also scale-free behavior hidden in the logistic equation.[8] To uncover this scale-free behavior we'll start by thinking about the transition from period 1 to period 2. In the upper-left plot of Fig. 7.16 I have plotted with a dotted line the logistic equation for the parameter value $r = 2$. I'll denote this function as $f_1(x)$. For this parameter, the orbits are drawn toward a period-one attractor—that is, a fixed point. The $y = x$ line is shown in gray. The attracting fixed point occurs at $x = 0.5$, where the logistic equation crosses the $y = x$ line. I chose $r = 2$ because for this parameter value the period-one attractor is most strongly attracting; among all possible r-values that give a period-one attractor, for $r = 2$ the rate of attraction is the largest. Such an attractor is called *super-stable* or *super-attractive*. See for example, Peitgen et al. (1992, pp. 596–8). So $f_1(x)$ is the logistic equation that has a super-stable fixed point.

Let's now turn our attention to period two. I'll choose to look at the logistic map for $r = 3.236$. This is the parameter value for which the period-two attractor is super stable. I'll call the logistic equation with a super-stable period-two attractor $f_2(x)$. The reason I'm focusing on super-stable behavior is as follows. I want to compare the period-one and period-two logistic equations. But the logistic equation has these properties not for a single r value, but for a range of r values. By always choosing the super-stable r values, I'm comparing "apples to apples."[9]

---

8. The discussion in the rest of this section closely follows Peitgen et al. (1992, Section 11.3) and Strogatz (2001, Section 10.7).

9. Also, one can show that the distances between super-stable r values in successive period doublings are the same as the distances between bifurcation points.

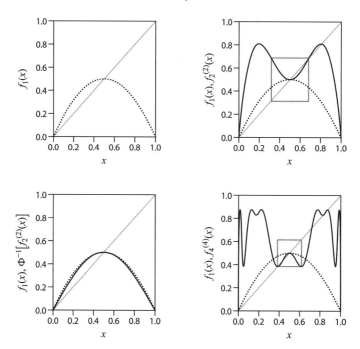

Figure 7.16. Graphs of the the logistic equation composed and scaled in different ways to show self-similarity. For further discussion, see text.

Now, if a point x^* is period-two for $f_2(x)$, this means that

$$f_2(f_2(x^*)) = x^* . \tag{7.14}$$

In words, apply f_2 twice to x^* and you end up with x^* again; this is what it means for a point to be period two. I can also think of "apply f_2 twice" as a single composite function. I'll denote this single, function by $f_2^{(2)}(x)$. Using this notation, I can write Eq. (7.14) as

$$f_2^{(2)}(x^*) = x^* . \tag{7.15}$$

This means that I could calculate δ by using the super-stable r values instead of the r values at which the bifurcations occur.

So if x^* is period-two for f_2, then it is a fixed point of $f_x^{(2)}$. This notation might look odd, but there's nothing deep going on here—Eqs. (7.14) and (7.15) are just two different ways of stating the criterion for points of period two.

I've plotted the composite function $f_2^{(2)}(x)$ as a solid line in the upper-right part of Fig. 7.16. I've also shown $f_1(x)$, again as a dotted line. Note that $f_2^{(2)}(x)$ intersects the gray $y = x$ line four times. Let's think about why. The intersections at $x = 0.5$ and 0.81 are the cycle of period two: $f_2(0.5) = 0.81$, and $f_2(0.81) = 0.5$. The composite function $f_2^{(2)}(x)$ also intersects the $y = x$ line at $x = 0$ and $x = 0.69$. These are both unstable fixed points: $f_2(0.69) = 0.69$ and $f_2(0) = 0$. If a point is fixed, it also has period two. For example, since $f_2(0.69) = 0.69$, if follows that $f_2(f_2(0.69)) = 0.69$, and thus $f_2^{(2)}(0.69) = 0.69$. Again, there's nothing deep or new going on here; we're looking at the composite function $f_2^{(2)}(x)$ and using it to visualize fixed points and points of period two.

Here is the key observation: the functions $f_1(x)$ and $f_2^{(2)}(x)$ appear similar. To see this, look in the upper-right plot in Fig. 7.16 and focus on the portion of the graph of $f_2^{(2)}(x)$ that is inside the gray box. It looks a lot like a small, upside-down version of $f_1(x)$. Let's see just how similar the two functions are. I'll take the portion of $f_2^{(2)}(x)$ that is inside the box, flip it upside down, and expand it so that the gray box is as large as the full graph. Let's call this rescaled function $\phi^{-1}[f_2^{(2)}(x)]$. (I'll explain why I'm calling this ϕ^{-1} and not ϕ.) The idea is that ϕ^{-1} is an operator that takes a function as input, flips it upside down, and enlarges it. In the lower left plot of Fig. 7.16 I show $\phi^{-1}[f_2^{(2)}(x)]$ as a solid line together with $f_1(x)$, plotted as a dotted line. Sure enough, the two functions look very similar. Ah-ah. We've found some self-similarity.

The story continues. I can do a similar thing with $f_4(x)$, the logistic equation with $r = 3.498$, which has a super-stable period-four attractor. Denote by $f_4^{(4)}(x)$ function f_4 applied four times: $f_4^{(4)}(x) = f_4(f_4(f_4(f_4(x))))$. The composite function $f_4^{(4)}(x)$

is plotted as a solid line in the lower-right graph on Fig. 7.16. The portion of $f_4^{(4)}(x)$ in the gray box closely resembles the portion of $f_2^{(2)}(x)$ in the gray box in the upper right figure, which also closely resembles $f_1(x)$. This self-similarity is what we are after. We have found self-similarity in $f(x)$, as anticipated.

I can now state what the renormalization operator ϕ is for this example: ϕ is an operator that takes a function as input and returns a re-scaled version. Geometrically, ϕ takes the unit square, flips it upside down, and shrinks it so it fits in the gray box in the upper right of Fig. 7.16. The result of applying this transformation to the function at some super-stable periodic parameter value will be a new function that is very similar to the function that has twice the period of the original function. For example, $\phi[f_1(x)] \approx f_2^{(2)}(x)$. The operator ϕ takes $f_1(x)$ and rescales it so that it looks like the portion of $f_2^{(2)}$ that is in the gray box in the upper-right part of Fig. 7.16.[10] We could then apply ϕ again to get a function that is $f_4^{(4)}(x)$. We are thus iterating ϕ. It turns out that ϕ has an attracting fixed function, just as many of the dynamical systems we have studied have a fixed point. The fixed function is a function that does not change when ϕ is applied to it. That is, $\phi[f_\infty] = f_\infty$. The function $f_\infty(x)$ can be thought of as the composite function for the r-value that has an infinitely long period—just before the transition to chaos.

This fixed function, which is usually called a *universal function*, is an attractor. This means that different initial conditions (i.e., functions) will all approach f_∞ when iterated with ϕ. We could start with the exponential function $f(x) = axe^{-x}$, or any other function with a quadratic maximum, and we would end up at the same universal function f_∞. Once we have an expression for ϕ and f_∞, it is possible to determine the exponent δ

---

10. The operator ϕ^{-1} moves in the other direction. It takes the function $f_2^{(2)}(x)$ and scales it so it resembles $f_1(x)$.

that describes the ratio of pitchfork lengths in the bifurcation diagram. Note that to calculate δ we do not need to know the initial condition—the particular function we start with—only ϕ and f_∞. The renormalization process thus explains why δ is universal: it is because the process of period doubling is moving one toward a universal function.

This discussion obviously isn't at all rigorous, but I hope it gives a flavor of what goes on in a renormalization calculation. Carrying this process out with algebra instead of graphs is quite involved. The Further Reading section at the end of this chapter has a number of suggestions for references at varying levels that you can turn to if you want to learn more.

7.7.3 Some General Remarks on Renormalization

I'll conclude this section with some general comments about renormalization. Renormalization is a standard technique in theoretical physics. Originally developed in high-energy field theory in the 1960s, it quickly found applications in statistical physics as well, and it's this form of renormalization that is used in the study of dynamical systems. Renormalization is a well understood theoretical line of attack on a bunch of interesting problems all concerning some sort of transition where there is scale-free or fractal behavior. The mathematics of renormalization can be rather involved, and my sense is that it is not a standard part of undergraduate curricula. That said, the math is not forbiddingly difficult; most elements of renormalization theory should be accessible to someone with a strong calculus background and a bit of fortitude.

In Section 7.6 I noted that there were two mysteries. The first was to figure out how universality could be possible for one-dimensional functions. Renormalization explains this mystery. The second mystery is how physical systems and higher-dimensional dynamical systems can have the same universal properties that we see in one-dimensional functions. Renormalization does not explain this mystery. Presumably there is some feature of higher-dimensional systems that is captured by one-dimensional iterated functions. We will see in Sections 9.4 and 9.5 that this is indeed the case.

7.8 Phase Transitions, Critical Phenomena, and Power Laws

The period-doubling transition to chaos can be viewed as a phase transition. So in this section I'll take a bit of a detour and offer some comments on phase transitions. The idea of a phase transition comes up often in some discussions of complex systems, and in my experience is a term that is occasionally used imprecisely. My aim is to untangle a handful of terms: phase transitions, critical phenomena, and power laws, as well as bifurcations and tipping points, which were discussed in Section 6.4.

First, a reminder that a bifurcation is a sudden, qualitative change in the number and/or stability of the fixed points of a dynamical system as a parameter is continually varied. We discussed bifurcations in Chapter 6. Closely related to bifurcations are phase transitions. A *phase transition* is a sudden qualitative change in the properties of a thermodynamic system. By a thermodynamic system I mean a system that consists of a very large collection of interacting entities, typically molecules. Mathematically, the number of entities is usually taken to be infinite; in practice the number might be on the order of 10^{23}. So bifurcations and phase transitions are similar, but are terms that originated in different contexts: bifurcations for dynamical systems and phase transitions for thermodynamic systems.

One approach to studying phase transitions leads to mathematics that is very similar to that which we encountered in our treatment of bifurcations in dynamical systems. One can choose to ignore statistical fluctuations in a thermodynamic system and instead focus on the average behavior of a salient quantity. This approach is known as *mean field theory*, since in ignoring fluctuations we are making the assumption that each degree of freedom in the system interacts with the average, or mean, field produced by all the other degrees of freedom. In mean field theory the system is described by just one average variable. One then can solve for the equilibrium state of the system by finding the minimum of a thermodynamic potential such as the Helmholz or Gibbs free energy plotted as a function of the mean variable. The shape of the free energy function changes as a parameter such as temperature is varied. These variations can lead to a sudden change in the minima of the thermodynamic system, just as changing a parameter in a differential equation can lead to a sudden change in the number of zeros of the right-hand side of the differential equation, as we saw in Figs. 6.2 and 6.8. So the mean-field analysis of phase transitions essentially reduces to the same mathematics that describes bifurcations in differential equations. Mean field theory is a standard topic in most statistical mechanics texts; Chapter 4 of Yeomans (1992) is a particularly clear introduction.

The phase transition we are probably most familiar with is the freezing or boiling of water. As one cools down liquid water, it transitions from liquid to solid at 0 C. The transition is not a gradual one. As we cool it, the water does not get more and more syrupy and viscous and gradually harden into ice. Rather, the water suddenly goes from liquid to a solid; there is nothing in between. One says that water transitions from the liquid phase to the solid phase; these two phases have qualitatively different properties.

The period-doubling route to chaos can also be viewed as a phase transition. For the period-doubling route to chaos, the two "phases" are periodic and chaotic. (I put phases in quotes in the previous sentence because one doesn't usually refer to periodic and chaotic regimes as phases.) The transition point for the logistic equation is $r_c \approx 3.56995$. Above this parameter value the dynamical system is chaotic; below this value it is periodic.

7.8.1 Critical Phenomena

There are two types of phase transitions: *discontinuous* and *continuous*. These types of phase transitions are also commonly called *first-order* and *second-order*, respectively. The discontinuous/continuous designation is a bit more modern and commonly used than first/second-order. In discontinuous phase transitions, as the name suggests, a quantity changes discontinuously. For example, in the liquid-solid transition of water, the density changes discontinuously;[11] liquid water has a density of $1g/cm^3$, while solid water (also known as ice) has a density of $0.92g/cm^3$. First-order phase transitions are not universal in the way that the period-doubling route to chaos is.

Continuous phase transitions are those in which a quantity of interest changes suddenly but continuously. For example, in a magnet–non-magnet transition, a material's magnetization rises from zero suddenly as the temperature is decreased. The temperature value at which this transition occurs is known as the critical point. Near the critical point, the system exhibits fractal-like behavior; there are fluctuations at all length scales and there are long-scale correlations. Mathematically, this fractal-like behavior manifests itself as a power law[12] distribution of fluctuations. Functions such as the specific heat or the magnetic susceptibility that measure the system's response to an external perturbation

11. In physical systems, the discontinuity associated with first-order phase transitions gives rise to a latent heat. For example, in order to solidify a kilogram of water, 330,000 J of energy must be removed it. This amount of energy is known as the latent heat of solidification or the latent heat of freezing.

12. A discrete random variable X is power-law distributed if the probability that X takes on the particular value x is proportional to $1/x$ raised to some power. That is,

$$P(X = x) = Cx^{-\alpha}, \tag{7.16}$$

where C is a normalization constant. Power-law distributions have *long tails*—they decay much more slowly than an exponential or Gaussian distribution. Thus, the occurrence of extreme values is much more likely for a power-law distributed variable than more typical exponential or Gaussian variables.

often diverge as a power law near the critical point. The behaviors associated with continuous phase transitions are collectively referred to as *critical phenomena*. The exponents on the various power laws that are observed near the critical point are called *critical exponents*.

Continuous phase transitions are interesting, rich phenomena that have attracted considerable attention from physicists and others. One of the reasons for this interest is that a continuous phase transition involves some form of collective behavior. In such a transition, many degrees of freedom that interact locally conspire to produce long-range correlations—correlations between system components decrease as a power law as their separation increases. The system's behavior is scale free; there are fluctuations on all length scales. Moreover, the exponents that describe the various power laws that one observes are universal in the same way that δ is universal for the period-doubling route to chaos.

Continuous phase transitions are grouped into a handful of *universality classes*. Which universality class a continuous phase transition belongs to is determined by the dimension of the system and the symmetry properties of the quantity that experiences a sharp change. (The technical name for such a quantity is an *order parameter*.) By symmetry I mean whether the quantity is a scalar, a vector in two-dimensional space, a vector in three-dimensional space, and so on. A given universality class will contain many different systems, and all systems in the same class will have the same critical exponents. In the same way, all iterated quadratic functions that undergo the period-doubling route to chaos will have the same value of δ.

7.8.2 Musings on Power Laws

As previously noted, at or near the critical point, a system that undergoes a continuous phase transition is almost always characterized by a power law distribution of some sort. However, the converse is resoundingly not true: if one observes a power law, that does not mean that the system is at or near a critical state. This is an important point, and something that people very often get confused about, so I'll say it again. Observing a power law does not mean that one is at a critical point—that the system is poised between two phases. Nor does it mean that the system is in a state characterized by organization or some sort of collective behavior. There are many processes that produce power laws that have nothing to do whatsoever with critical phenomena and are not associated with complexity or memory. There have been some excellent essays and review papers on these issues over the last decade or so: Reed and Hughes (2002), Mitzenmacher (2004), Newman (2005), Keller (2005), Stumpf and Porter (2012). Readers wanting to know more about the diversity of mechanisms that can produce power-law data are encouraged to check these references out. I review some of these mechanisms in Unit 6 of my online course on fractals and scaling (Feldman, 2015). See also chapters 16 and 17 of Mitchell (2009) for an even-handed look at the roles and relevance of power laws for complex systems. There are several pitfalls associated with detecting the existence of power laws in empirical data and accurately and reliably estimating their exponents. For a discussion of these pitfalls and a careful explanation of how to avoid them, see Clauset et al. (2009).

Also, it is important to remember that not all phase transitions are critical phenomena. Many, if not most, phase transitions are discontinuous, and so are not universal and are not characterized by power laws or diverging correlations. In my experience the distinction between continuous and discontinuous phase transitions is occasionally misunderstood, and some imply that all transitions, regardless of their type, are associated with power laws and universal behavior.

7.8.3 Exporting Physics Terminology

The study of phase transitions began in physics. And within physics, the term phase transition has a fairly precise meaning: a discontinuous or non-smooth change in the properties of a thermodynamic system. But the concept of a phase transition has been exported from physics and now is often used to describe any relatively sudden change in a system, thermodynamic or not, including those transitions for which the change is not infinitely sharp. Similarly, the term critical phenomena, which within physics would describe only continuous phase transitions, is also sometimes used to describe other transitions, both first-order phase transitions and transitions that are, strictly speaking, not phase transitions at all, since they are not infinitely sharp or sudden.

Although I occasionally find this slightly imprecise use of physics terminology to be troublesome, overall I think this is a reasonable state of affairs. It is indeed the case that our world is one in which properties of systems change abruptly: epidemics suddenly arise and spread, stock markets crash, species become extinct, empires fall, and social norms can change in less than a generation. Calling these sudden changes phase transitions seems not unreasonable. What I think is less reasonable, however, is to assume that these transitions share common features or are universal, just because some physics phase transitions are universal. It may indeed be the case that there are some common mechanisms among transitions in biological and social systems, but this is a proposition that requires support—it does not follow automatically from the simple observation that there exist sudden transitions.

The term "universal" has also come to be used outside of physics. And as with the term "phase transition," the non-physics use of universal is broader and a bit fuzzier than the original use of the term within physics. As Stein and Newman note (2013, p. 282), "over the years the meaning [of universality class] has broadened to a less precise notion of a class of systems that display the same general behavior despite being very different in nature."

I'll conclude this section by mentioning a few more ideas related to phase transitions. In ecology and earth science the term *regime shift* is used to describe a transition from one type of dynamical behavior to another. For example, a region may transition from prairie to desert, or a lake may shift from a turbid state to one which is dominated by vegetation. Transitions of this sort are clearly similar to bifurcations or phase transitions. A similar notion is that of a *tipping point*, taken to be a threshold at which a transition occurs, as briefly discussed in Section 6.4.

Recently there has been interest in *critical transitions*, defined by Scheffer (2009, p. 104) as "the subclass of regime shifts that in models could correspond to... transitions in which a positive feedback pushes a runaway change to a contrasting state once a threshold is passed." The basic idea seems to be that in a critical transition, if one crosses the transition point, one is pulled very quickly to an alternative attractor. There has been much work on developing methods to detect critical transitions. (See e.g., Dakos et al. (2012) and Scheffer et al. (2012)). Some argue that near the transition point one might observe slower dynamics and larger fluctuations, as is the case with critical phenomena in physical systems. My sense is that this is an ongoing area of research, and while there have been some promising results for model systems, detecting critical transitions using only data has, unsurprisingly, proved more difficult.

7.9 Conclusion: Lessons and Limits to Universality

What are the take-away messages from universality? What are its implications for the study of complex systems? My thoughts on these questions are mixed. To start, I think universality is unarguably a stunning result for mathematics and physics.

A remarkably simple equation, the logistic equation, has made accurate, quantitative predictions about a range of physical phenomena that seem completely unrelated to whatever caricature of population dynamics that the logistic equation was intended to capture. Even though renormalization provides a convincing mathematical understanding of how this can be so, it still seems to me to be a bit magical.

Universality tells us that the period-doubling route to chaos in almost all phenomena are characterized by a δ of 4.669. This fact lets us make quantitative statements about some systems without detailed knowledge of their physics. For example, one phenomenon that undergoes the period-doubling transition is convection rolls in a fluid. When heated from below in a container of the appropriate size, fluid will form two convection rolls. As the temperature is increased, the rolls develop a simple wiggle at a certain frequency. As the heat is increased further, the rolls develop more complicated wiggles—first with period two, and then period four, and so on.

The key parameter in this experiment is the difference between the temperatures of the top and bottom of the container, which I'll call ΔT.[13] Suppose that the wiggle in the rolls first appears when ΔT is 1 degree, and that the period two wiggles appear at a ΔT of 2 degrees. Then you can determine approximate values for

---

13. Results for these experiments, e.g., Libchaber et al. (1982), are usually expressed in terms of the Reynolds number, which is a dimensionless quantity used in fluid dynamics to characterize fluid properties. In the context of these experiments, Reynolds number is directly related to ΔT, the temperature difference between the top and bottom of the container. I will tell the story using temperature, because this is a more familiar quantity than Reynolds number. Also, in order to keep the numbers in my example simple, I've deliberately chosen unrealistic temperature values; in actual experiments, ΔT is several milli Kelvin.

By the way, the original experiments I described here were done by Albert Libchaber in the 1980s (Libchaber et al. (1982); Libchaber and Maurer (1982)). These experiments were fiercely difficult; see the chapter titled "The Experimenter" in Gleick (1987). An interesting, short profile of Libchaber is Mukerjee (1996).

the ΔT's at which all the subsequent period doublings occur. For example, the next period-doubling occurs when $\Delta T \approx 2.214^{14}$. What's noteworthy about this is that you can figure this out without knowing *anything* about fluid mechanics. Since the period-doubling route to chaos is universal, it does not depend on details of the system.

The discovery and appreciation of universality is an example of anti-reductionist science. In broad strokes, *reductionism* is the belief that the best line of attack on a scientific problem is to try to understand something by learning about its constituent parts. It is surely the case that learning about something's parts can be very instructive. We've certainly learned a lot about biology by understanding cells and genes. And we've learned a lot about cells and genes by understanding organic chemistry, which in turn we can learn a lot about by studying quantum mechanics. But the study of dynamical systems often takes us in the other direction, looking for general dynamical mechanisms and phenomena that are common across many systems—seeking larger patterns rather than smaller parts. I think that universality in chaos is among the most successful anti-reductionist examples; it successfully predicts quantitative features of the transition to chaos in starkly different physical systems.

In Sections 3.4 and 3.5 we discussed the merits of using simple models—models that are designed to serve as a caricature or sketch of some key features of the system under study. Such models were presented as a stylistic or subjective choice. The assertion or hope is that these simple models give us some insight into the more complex systems that they are designed to model. The phenomenon of universality provides mathematical and experimental vindication of this choice. It tells us that it is indeed the case that for some

---

14. To see this, note that

$$\frac{2-1}{2.214-2} \approx 4.67 \,.$$

classes of systems certain features—the critical exponents—are largely independent of the details of the system.

So what does universality hold for the study of complex biological and social systems? Is there any reason to expect to observe universal phenomena? My fairly strong hunch is probably not. I think it is clear that complex systems sometimes undergo transitions from one type of behavior to another. And, to varying degrees, these transitions can be viewed as phase transitions. But critical phase transitions are relatively rare, even in the physical realm. I can think of no critical phase transition that is commonly experienced in everyday life; the freezing and boiling of water are both discontinuous transitions, and hence not universal. Also, the presence of quantities distributed according to a power law is not sufficient to conclude that one is observing a critical phenomenon. Power laws are common in complex systems, but there are many ways to generate power-law behavior that have nothing to do with critical phenomenon and universality.

In fact, I suspect that the discovery and understanding of universality in transitions to chaos and continuous phase transitions may actually be a somewhat misleading guide for the study of complex systems. I think it is extremely unlikely that one will find quantitative universality in the study of complex systems. There may be qualitative universality—generic mechanisms of observed phenomena. But I think that the search for critical phenomena and quantitative universality in complex systems may be a blind alley. A notable exception could be the West–Brown–Enquist theory of metabolic scaling (West et al., 1997; West and Brown, 2004; West, 2017), which predicts and explains a scaling law that describes the metabolic rates of organisms from amoebas to whales. It is important to note, however, that this relationship, which is often called universal, does not arise from a critical phenomenon or phase transition.

I should be clear, though, that these comments are my opinions and are not universally shared. Some argue that biological and perhaps even social dynamics can drive (though natural selection or other reward mechanisms) a system to tune its parameters so that it is at or near a critical point. The argument is that being near a critical point is advantageous because a critical system has fluctuations at many scales, and thus is better suited to adapt or respond to a changing environment. (See, e.g., Mora and Bialek (2011), Ball (2014), and Attanasi et al. (2014)).

7.10 Further Reading

There are a wide range of accounts of universality and renormalization. Popular discussions can be found in Stewart (2002) and Gleick (1987). Feigenbaum (1983) wrote a semi-technical account of his discovery and analysis of universality. Thesis 4 of Aubin and Dahan Dalmedico (2002) is a thorough and nuanced historical analysis of universality. See also Mitchell (2009, pp. 34–38), the recent history by Coullet and Pomeau (2016), and "The Butterfly Effect" by Ghys (2015). A short general-audience discussion of critical phenomena in physics and networks is Watts (2004, pp. 61-68).

A technical but accessible treatment of the period-doubling route to chaos, including a discussion of universality, is Peitgen et al. (1992, Chapter 11). See also Chapter 6 of Smith (1998). A much more formal overview of period-doubling is Tresser et al. (2014). The edited volume by Cvitanović (1989) contains many of the early papers on universality in dynamical systems, including experimental observations of universality.

Discussions of the renormalization group applied to dynamical systems are, by necessity, quite technical. A particularly clear reference is (Strogatz, 2001, Section 10.7). See also Coppersmith (1999), Hilborn (2002, Chapter 5 and Appendix F), and Schuster and Just (2006, Chapter 4). An excellent introduction to the applications of the ideas and techniques of renormalization to complex systems is the Complexity Explorer Course by DeDeo (2016).

8

HIGHER-DIMENSIONAL SYSTEMS AND PHASE SPACE

Thus far in this book I've focused exclusively on one-dimensional dynamical systems. We've looked at how a single number—a population or temperature or whatever—changes over time. In this chapter I'll introduce a few higher-dimensional systems and present the idea of *phase space*—a powerful geometric technique for visualizing and thinking about the behavior of dynamical systems. In the subsequent chapter, we'll look at how chaos manifests itself in higher dimensions, a highlight of which is strange attractors. I'll begin by quickly reviewing one-dimensional differential equations that I first presented in Chapter 2.

8.1 A Quick Review of One-Dimensional Differential Equations

In Chapter 2 we looked at differential equations of the following form:

$$\frac{dP}{dt} = f(P) . \tag{8.1}$$

I'm using the variable P here because I'm thinking of the variable as representing a population that varies in time. Equation (8.1) tells us how the growth rate of the population (dP/dt) depends on the population P. This dynamical system is one-dimensional, because there is only one variable, the population P, whose behavior we are interested in.

In Section 6.1 we looked at the logistic differential equation:

$$\frac{dP}{dt} = 3P \left(1 - \frac{P}{100}\right), \qquad (8.2)$$

a differential equation that is of the form of Eq. (8.1). This equation describes the growth of a population that has a carrying capacity of 100. We analyzed this equation graphically in Fig. 6.1. I've reproduced these plots here in Fig. 8.1 for easy reference. In the top plot in the figure we see the right-hand side of Eq. (8.2)—that is, a plot of the growth rate versus the population P. We see that the population grows (dP/dt is positive) if P is less than 100 and the population shrinks (dP/dt is negative) if P is greater than 100.

There are two aspects of Fig. 8.1 that I want to remind you of before we move to considering two-dimensional systems. First, recall that the solution to a differential equation of the form of Eq. (8.1) is $P(t)$: the population as a function of time. In order to figure out $P(t)$, we need to know the initial value of the population; all future values of the population are determined by the differential equation. In the bottom part of Fig. 8.1 I have shown four different solutions $P(t)$ to the differential equation.

The second thing I want remind you about is that we can summarize the behavior of all solutions to a differential equation of the form of Eq. (8.1) with a *phase line*. I've shown the phase line for Eq. (8.2) on the horizontal axis of the top plot in Fig. 8.1. The phase line shows that all populations greater than 0 will move toward the stable equilibrium at $P = 100$. The phase line tells us which way solutions move at any point on the line. This lets us see quite clearly the global behavior of the differential equations—the fate of any initial condition. What the phase line does not tell us, however, is how fast the population will move along the phase line. To observe this, we would need the solution $P(t)$.

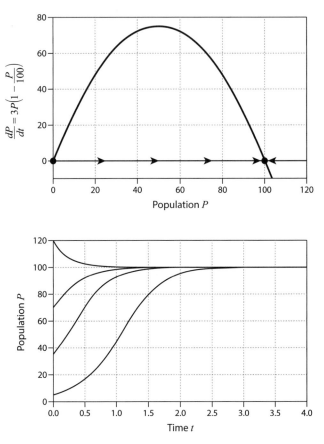

Figure 8.1. Top: A plot of the right-hand side of Eq. (8.2) and the phase line for the equation. There is a stable fixed point at $P = 100$ and an unstable fixed point at $P = 0$. Bottom: The solutions $P(t)$ to Eq. (8.2) for four different initial conditions: $5, 35, 70$, and 120. Note that all solutions approach the stable equilibrium at $P = 100$. This figure is identical to Fig. 6.1.

8.2 Lotka–Volterra Differential Equations

Having reviewed one-dimensional differential equations, we now turn our attention to two-dimensional differential equations. I'll do so via a simple and well known model of two interacting species, the Lotka–Volterra (LV) model. The model considers two different populations, one prey, and the other a predator. For concreteness, I'll use rabbits and foxes, and denote their populations by R and F, respectively. Rabbits are the prey; they get eaten by the foxes.

The growth rate of the rabbits (dR/dt) depends on both the current number of rabbits R and the current number of foxes F. The same is true for the growth rate of the foxes (dF/dt), although the functional dependence need not be the same. The general form for a differential equation describing this situation is:

$$\frac{dR}{dt} = f(R, F),$$

$$\frac{dF}{dt} = g(R, F). \qquad (8.3)$$

Compare these equations with Eq. (8.1), where the growth rate dP/dt of the population is a function of the current population P. In Eq. (8.3) there are two populations, R and F, each of whose growth rates, dR/dt and dF/dt, depend on both R and F. One says that Eq. (8.3) is a system of two *coupled ordinary differential equations*.

The Lotka–Volterra model consists of a particular form of the coupling functions $f(R, F)$ and $g(R, F)$. The general form for the LV model is:

$$\frac{dR}{dt} = aR - bRF,$$

$$\frac{dF}{dt} = cRF - dF, \qquad (8.4)$$

where a, b, c, and d are model parameters, all of which are positive. The parameters have the following interpretations: a is the growth rate of the rabbits in the absence of foxes, b is a measure of how deadly foxes are to the rabbits, c is a measure of how beneficial rabbits are to the foxes, and d is the death rate of the foxes. The minus sign in front of the parameter b means that the presence of foxes causes the rabbit population to decrease. Similarly, the presence of rabbits causes the fox population to increase, since the cRF term is positive. Since R and F both represent populations, we will restrict our analysis of Eq. (8.4) to non-negative values of R and F. I will not give a full motivation of the LV model here, as it is a standard topic and is well-treated elsewhere.[1] Our interest here is to study Eq. (8.4) as a dynamical system. How might we produce solutions and what are their long-term behaviors?

In solving the logistic differential equation, Eq. (8.1), we looked for the population P as a function of time. That is, we sought $P(t)$. Here we are looking at *two* populations: the rabbits and the foxes. So we seek $R(t)$ and $F(t)$. We solved Eq. (8.1) (i.e., we found $P(t)$) by using Euler's method, discussed in Section 2.5. Recall that when using Euler's method we pretend that the ever-changing rate of change dP/dt is constant for small intervals Δt and then use this pretend-constant rate of change to figure out the change in the population over each time interval Δt. An essentially identical procedure works for two-dimensional systems like Eq. (8.3).[2]

So let's look at some solutions to the Lotka–Volterra equations obtained via Euler's method. To find such solutions, I need to choose numerical values for the four parameters in Eq. (8.4). I chose $a = 1$, $b = 0.25$, $c = 0.2$, and $d = 0.6$. I chose these numbers for the sake of convenience—I'm not claiming that these numbers, or the LV equations themselves, necessarily say anything true about actual rabbits and foxes. To come up with a solution to the differential equation I also need to choose initial conditions:

---

1. Derivations can be found in many textbooks on differential equations and mathematical modeling, e.g., Edelstein-Keshet (2005, Secs. 6.2 and 6.3) or Feldman (2012, Chapter 30). I sketch a derivation of the LV equations in Unit 7.1, Video 3 of my online course on chaos and dynamical systems (Feldman, 2014). The Wikipedia article on the Lotka–Volterra equations is also good (Wikipedia, 2016).

2. I won't go through the calculation in detail. It is conceptually no different than Euler's method in one dimension. For a more thorough treatment, see, e.g., Feldman (2012, Chapter 30).

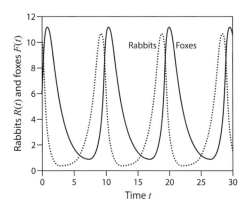

Figure 8.2. Solutions to the LV system, Eq. (8.4), obtained via Euler's method. The parameters are: $a = 1$, $b = 0.25$, $c = 0.2$, and $d = 0.6$. The initial conditions are $R(0) = 10$ and $F(0) = 6$. The solution to the system of differential equations consists of two functions: $R(t)$ and $F(t)$, the values of the rabbits and the foxes as a function of time.

starting values for the rabbits and the foxes. I chose $R(0) = 10$ and $F(0) = 6$. Again, I chose these for numerical convenience, not biological fidelity.

The solutions I obtained via Euler's method are shown in Fig. 8.2. We see that both populations oscillate. One can tell a simple story to explain these oscillations. Initially the fox population increases and the rabbit population decreases. The foxes are benefiting by eating the rabbits, and their population goes up, while the rabbits suffer from being eaten, and their population goes down. But as this continues, the foxes start to suffer; there are not enough rabbits for them to eat. From roughly time 1 to time 7, the fox population is decreasing. By around time 3, the rabbit population begins to recover, since there are fewer foxes around to eat them. Then, at $t \approx 7$, the fox population begins to increase again, since there are now plenty of rabbits to eat. And the cycle continues. The foxes eat the rabbits, the rabbit population decreases, then the fox population decreases, then the rabbit population recovers, then the fox population recovers.

Figure 8.2 shows that two-dimensional systems of differential equations are capable of more types of behavior than one-dimensional differential equations. In Section 2.10 I argued that, due to determinism, solutions to a one-dimensional autonomous[3] differential equation cannot exhibit oscillations of any kind. A particular solution to such a differential equation can either increase or decrease (or remain constant), but not both. In terms of the phase line, the effect of the determinism of the differential equation is to make the phase line a one-way street. At a given point on the phase line, all solutions will either move to the left or to the right (or remain constant). But it is not possible that at a single point on the phase line there are two solutions moving in different directions.

8.3 The Phase Plane

The phase line gives us a global view of the dynamics of a one-dimensional differential equation: it lets us see the fate of all initial conditions. The phase line also lets us think about the types of dynamical behaviors that are—and aren't—possible, as I did in the previous paragraph where I argued against the possibility of oscillating solutions. There is a similarly useful geometrical construction for two-dimensional differential equations: a *phase plane* as opposed to a phase line.

Figure 8.3 shows the phase plane representation of the solutions to the LV equations that I plotted in Fig. 8.2. To make the phase plane plot, I plotted F against R. This allows us to see directly the relationship between F and R. The rabbit and fox populations move around the oval in a counter-clockwise direction. Figures 8.2 and 8.3 are two complementary ways of representing the solutions to the LV equation. You might want to take a moment, perhaps making a quick sketch, to see how to go from Fig. 8.2 to Fig. 8.3.

---

3. Recall that *autonomous* means that the right-hand side of the differential equation $dP/dt = f(P)$ has no explicit time dependence.

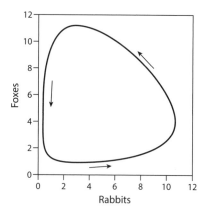

Figure 8.3. The solutions of Fig. 8.2 plotted in the phase plane.

Note that it is not possible, however, to go in the other direction; we cannot determine the solutions $R(t)$ and $F(t)$ from their phase plane plot in Fig. 8.3. The reason for this is that there is no time dependence on a phase plane. Figure 8.3 tells us that the populations move along the oval in a counter-clockwise direction, but we don't know how fast they traverse the oval. The phase plane plot indicates that $R(t)$ and $F(t)$ will each oscillate, but we can't determine how long each oscillation takes.

To help make sense of Fig. 8.3, it may be helpful to use it to tell the story of the intertwined fates of the rabbits and foxes. Let's start in the lower left corner of the oval. Here there are few foxes and few rabbits. The rabbit population then grows, unchecked by hungry foxes, and the solution moves to the right—the direction of increasing rabbits. But gradually, as the rabbit population increases, the fox population increases too. Then the rabbit population is around 10.5—the right-hand "corner" of the oval—the rabbit population begins to decrease, suffering losses due to the foxes. Meanwhile, the well-fed fox population grows. This continues until the fox population is roughly 11—the top "corner" of the oval. At this point the rabbit population is too small to support so many foxes. So the fox population decreases significantly and we

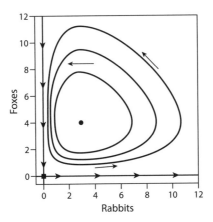

Figure 8.4. The phase portrait for the Lotka–Volterra system.

end up where we started, in the lower left part of the cycle with few foxes and few rabbits. This is, of course, the same story I told about Fig. 8.2.

To complete the Lotka–Volterra example we should think about other solutions. What happens for other initial conditions? Is the cycle shown in Fig. 8.3 attracting? To answer these questions, I've plotted some additional solutions to the LV equations on the phase plane in Fig. 8.4. The three ovals are the result of choosing three different initial rabbit and fox populations. For all ovals the initial fox population was $F(0) = 6$. The initial rabbit populations are $R(0) = 10$, 8, and 6. The outermost oval is the one shown previously, in Fig. 8.3. These ovals are neither repelling nor attracting. If the population is cycling along one oval and the number of foxes or rabbits changes a little, then the trajectory in the phase plane will move neither back to, nor away from, the original oval. Rather, the system will proceed along a new oval, more or less concentric to the original one.

The LV system has two equilibrium points. Recall that an equilibrium for a differential equation is a point at which the derivative is zero, and thus solutions to the differential equation are constant.

For a two-dimensional system an equilibrium (i.e., a constant solution) occurs when *both* derivatives are zero:

$$\frac{dR}{dt} = 0,$$

$$\frac{dF}{dt} = 0. \tag{8.5}$$

So to find the equilibrium values for the LV system, I need to set the right-hand sides of both parts of Eq. (8.4) equal to zero and solve for R and F. One solution is immediately apparent: $R = 0$ and $F = 0$. Biologically, this makes sense. If there are no rabbits and foxes now, there will be no rabbits and foxes forever. This equilibrium value is shown as a black square on Fig. 8.4.

There is another equilibrium point that is a bit less obvious, both algebraically and biologically. It turns that there is another set of R, F values that make the right-hand side of Eq. (8.4) zero:

$$R = \frac{a}{b},$$

$$F = \frac{c}{d}. \tag{8.6}$$

You can verify that if you plug these values for R and F into Eq. (8.4), that both derivatives really are zero. For the particular parameter values I've been using in this chapter, these equilibrium values are $R = 3$ and $F = 4$. If there are initially 3 rabbits and 4 foxes[4] then the two populations will remain at this value forever. I've indicated this equilibrium value on Fig. 8.4 with a black circle.

To conclude our analysis of the LV system, there are two more cases we need to think about. First, what happens if we start with no foxes but some rabbits? Plugging $F = 0$ into Eq. (8.4),

---

4. Remember that we're thinking of R and F in arbitrary units. So $R = 3$ might mean 30000 rabbits or 3000 kg of rabbits.

we have:

$$\frac{dR}{dt} = aR,$$

$$\frac{dF}{dt} = 0.$$ (8.7)

So we see that the fox population will remain at zero; this makes sense, because if there are no foxes in the world today, there will be no foxes in the world tomorrow. And Eq. (8.7) tells us that the rabbit population will grow exponentially. Its growth rate dR/dt is always positive, since there are no foxes to keep the rabbits in check. This scenario is illustrated on the horizontal axis of Fig. 8.4. Second, we need to consider what happens if there are no rabbits. Plugging $R = 0$ into Eq. (8.4), one gets:

$$\frac{dR}{dt} = 0,$$

$$\frac{dF}{dt} = -dF.$$ (8.8)

These equations tell us that the number of rabbits remains at zero, and the foxes, who have nothing to eat because there are no rabbits, decay to zero. This is shown on the vertical axis of Fig. 8.4.

Figure 8.4 lets us determine the behavior of any initial condition. If R and F are both non-zero, the resulting trajectory in the phase plane will be a counter-clockwise oval. The one exception is if R and F start right on the equilibrium value, $R = 3$, $F = 4$. Then the population is fixed; it will remain at these values. If R is zero but F is not, the system moves down the horizontal line on the left toward the origin. All the foxes die. And if F is zero but R is not, the system moves horizontally to the right, and rabbits take over the world. Figure 8.4 is known as a *phase portrait*: a phase plane with enough trajectories included so as to illustrate the fate of all initial conditions. A phase portrait is a phase space plot consisting of several trajectories and all equilibria so as to produce a clear view of the long-term fate of all initial conditions.

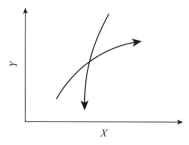

Figure 8.5. Two trajectories on a phase plane. The situation shown is not possible, because at the point where the trajectories cross, determinism is violated.

8.4 Phase Planes in General

Let's step back and think about the phase plane a bit more generally for equations of the form:

$$\frac{dX}{dt} = f(X, Y),$$

$$\frac{dY}{dt} = g(X, Y). \tag{8.9}$$

This is the same form as Eq. (8.3), but now I am using X and Y for the variable names instead of R and F, since we are not necessarily thinking about populations of rabbits and foxes. At each point on the phase plane, the differential equation specifies the direction that the solution will move. This is just a re-statement of what a system of two differential equations like Eq. (8.9) means—it is set of instructions that specifies the instantaneous rate of change of two variables, and these rates of change depend only on the current values of the two variables.

A consequence of this fact is that solutions on a phase plane cannot cross. To see this, consider Fig. 8.5, on which I have sketched two possible solutions to a system of two differential equations. At the crossing point the two curves are moving in different directions. But this isn't possible, because it violates the determinism inherent in Eq. (8.9)—the rates of change of X and Y (and hence their directions on the phase plane) depend only on the value of X and Y. That is, the direction of a solution trajectory is uniquely determined by its position on the phase plane. We thus conclude that solutions to Eq. (8.9) cannot cross on the phase plane.[5]

So trajectories in the phase plane cannot cross. This limits the types of behavior possible for two-dimensional differential equations. In particular, it is not possible to have a bounded, aperiodic trajectory. Thus, chaos is not possible for two-dimensional differential equations. Proving this result, known as the *Poincaré–Bendixson Theorem*, is beyond the scope of this book.[6] However, it is not hard to see why bounded aperiodic orbits are forbidden. As a trajectory moves through the phase plane, it continually boxes itself in, extending the line that it cannot cross. The result is that its available space effectively shrinks down to nothing, making it impossible for the trajectory to continue indefinitely. Thus, the only possible behaviors for dynamical systems of the form of Eq. (8.9) are:

1. Orbits can tend toward infinity—move infinitely far from the origin. This was the case if $F(0) = 0$ on the phase portrait of Fig. 8.4, where the rabbit population grows without bound.

2. Orbits can approach a fixed point. We saw this behavior if initially there are no rabbits in Fig. 8.4; the fox population approaches the fixed point at $(R = 0, F = 0)$.

3. Orbits can be periodic or approach a periodic cycle. We also observed this behavior in the Lotka–Volterra

---

5. This assumes, of course, that $f(X, Y)$ and $g(X, Y)$ are both well-behaved functions: deterministic, finite, and differentiable.

6. For a more technical discussion and proof of the Poincaré–Bendixson see, e.g., Section 10.5 of Hirsch et al. (2004) or Section 10.3 of Robinson (2012).

system in Fig. 8.4. There all orbits that start with both
R and *F* positive are immediately cyclic. For other
systems, it is often the case that an orbit is not exactly
cyclic immediately, but gets pulled toward a periodic
cycle. Attracting periodic cycles of this sort are called
limit cycles. A limit cycle is any closed, attracting
trajectory on a phase plane.

And that's it. That's all that two-dimensional autonomous differential equations can do.

What we're seeing is that the geometry of space in which the
solutions live—a one-dimensional phase line or a two-dimensional
phase plane—together with the determinism of the equations,
limits the possible dynamical behaviors. In one dimension, solutions cannot oscillate. In two dimensions, oscillations are possible, but not aperiodic ones. As a consequence, chaotic solutions
(bounded, aperiodic trajectories) do not exist in two dimensional
systems. What about three-dimensional differential equations?
We shall see that the additional dimension in which their solutions
live means that they are capable of much more complex behavior.
In the next section I'll introduce a simple system of three differential equations that we will use in the next chapter when we'll look
at some of the fun and complex phenomena exhibited by three-
dimensional differential equations, including chaos and strange
attractors.

8.5 The Rössler Equations and Phase Space

In this section and in the next chapter, we'll examine in some
detail a system of three differential equations known as the *Rössler
equations.* The equations are:

$$\frac{dx}{dt} = -y - z\,,$$

$$\frac{dy}{dt} = x + ay \,,$$

$$\frac{dz}{dt} = b + z(x - c) \,, \qquad\qquad (8.10)$$

where a, b, and c are parameters. These equations were concocted by Otto Rössler to be a simple system that has a strange attractor that is easy to visualize. (We'll encounter this strange attractor in the next chapter.) This is a system of three, coupled, first-order differential equations. Each equation specifies the rate of change of one of the variables as some function of the other three. A solution to this differential equation consists of three functions $x(t)$, $y(t)$, and $z(t)$.

As with the two-dimensional Lotka–Volterra system, we find solutions to Eq. (8.10) using Euler's method. The results of doing so are shown in Fig. 8.6.[7] We can see that all three solutions oscillate. The oscillations initially grow but by the around $t = 50$, the system has reached a steady state of some sort; the oscillations do not appear to be changing. Different initial conditions lead to the same long-term behavior. The steady-state behavior is attracting; all nearby solutions are pulled towards it.

Just as we did for the Lotka–Volterra equation, we can plot these solutions on the phase plane—except now it's not a plane but a three-dimensional phase space, since there are three variables: x, y, and z. Figure 8.7 shows several views of $x(t)$, $y(t)$, and $z(t)$ plotted in phase space. The large, bottom plot shows a three-dimensional rendering of the solutions in phase space—the curve traced out by the x, y, z coordinates. The upper left plot shows a top view of the trajectory; I plotted x versus y and did not plot z. In this view one can see the trajectory spiraling out,

---

7. Actually, this isn't true. I used a more efficient Runge–Kutta method, which I mentioned briefly in Section 2.7. I find that for three-dimensional systems—especially chaotic systems as we'll encounter in the next chapter—Euler's method is not always reliable or efficient.

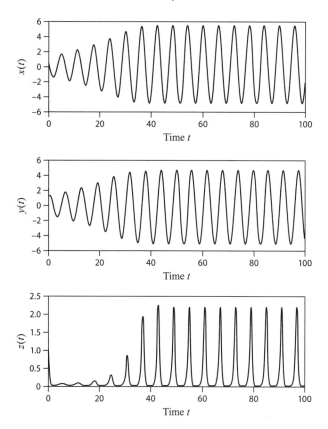

Figure 8.6. Solutions $x(t)$, $y(t)$, and $z(t)$ to the Rössler system, Eq. (8.10). The parameter values are $a = 0.1$, $b = 0.1$, and $c = 3.0$. The initial conditions are $x(0) = 1$, $y(0) = 1$, and $z(0) = 1.5$.

counter-clockwise, until it reaches the stable periodic cycle. This cycle is the slightly darker band. The upper right figure shows a side view of the same trajectory. I plotted x versus z and did not plot y.

To help see the periodic attractor more clearly, in Fig. 8.8 I have shown the trajectories without the transient behavior. I did this by making a phase space plot of x, y, and z from Fig. 8.6 using only the values from time 50 onwards. By this point the

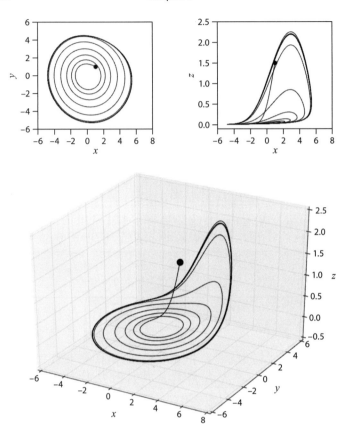

Figure 8.7. The solutions of Fig. 8.6 plotted in phase space. The top two plots show a top and side view. The bottom plot is a "three-dimensional" rendering. The darker, saddle-shaped loop is the periodic attractor. The starting point for the trajectory, $(x(0) = 1, y(0) = 1, z(0) = 1.5)$, is indicated with a black dot.

solutions are already on the attractor, and so the resulting phase space plot shows only the periodic behavior. Viewed from above, motion on the attractor is counter-clockwise. This shape is an attractor; nearby initial conditions will be pulled in toward this shape. Equivalently, we would say that this dynamical behavior is stable. If a trajectory is on the attractor and then is bumped off it, the trajectory will return to the attractor.

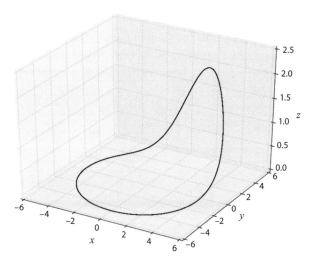

Figure 8.8. The periodic attractor for the Rössler system, Eq. (8.10), with $a = 0.1$, $b = 0.1$, and $c = 3.0$. This figure shows the same trajectories as shown in Figs. 8.6 and 8.7, but they are plotted from $t = 50$ to 100, not $t = 0$ to 100. Viewed from above, the motion on the attractor is counter-clockwise.

Note that trajectories in phase space in Fig. 8.7 do not cross—they can't, for the same reason that trajectories in one- and two-dimensional phase spaces can't: doing so would violate determinism. In Fig. 8.7 trajectories appear to cross, but they never actually touch; one is always above or behind the other. Three-dimensional dynamical systems have more space in which to operate. A trajectory in a three-dimensional phase space does not create an impassible barrier; the trajectory can pass under or over it. This hints that three-dimensional dynamical systems are capable of more complex behavior than two-dimensional dynamical systems. We will see in the next chapter that this is indeed the case.

To conclude, a note about terminology. Phase space is very often referred to as *state space*. State space is defined as the set of all possible states of a dynamical system. For the Rössler equation, the state space is all possible values of x, y, and z. For the LV system, the state space is all possible values of the populations R and F. Like many authors, I am using phase space to be synonymous with state space. Some authors, however, use phase space to refer only to a state space that forms a smooth manifold (e.g., Terman and Izhikevich (2008)) or to describe the state space for a Hamiltonian system (e.g., Hilborn (2002)). A Hamiltonian system is a dynamical system that conserves energy.

8.6 Further Reading

The discussion in Sections 8.2 and 8.3 give a minimal overview of the analysis of two-dimensional differential equations. This is a rich and interesting area of mathematics, and there are many standard and not-too-difficult analytic techniques that complement the numerical and experimental approach I've taken here. Most modern texts on differential equations treat two-dimensional differential equations in considerable detail. Of the books that I'm familiar with, I think Blanchard et al. (2011) and Hirsch et al. (2004) are particularly clear. Two-dimensional systems of differential equations are also covered in many books on modeling; I'd particularly recommend Edelstein-Keshet (2005).

The idea of phase space—plotting dynamical variables against themselves instead of time—has been a powerful and far-reaching abstraction. Note that, for the Lotka–Volterra equations, the phase plane ends up being a fairly abstract space. Lengths in this space are not measured in meters or inches, but in numbers of animals. Phase space is a construction that is very commonly used in applied mathematics and across the sciences, but this was not always so. A short history of phase space is the article, by Nolte, "The tangled tale of phase space" (2010). A succinct and clear overview of phase space is Terman and Izhikevich (2008). A discussion of references about the Rössler equations can be found at the end of Chapter 9.

9

STRANGE ATTRACTORS

In the previous chapter we looked at the Rössler equations, a system of three, coupled, autonomous, ordinary differential equations. The equations are:

$$\frac{dx}{dt} = -y - z,$$

$$\frac{dy}{dt} = x + ay,$$

$$\frac{dz}{dt} = b + z(x - c). \tag{9.1}$$

We saw that for a certain set of parameter values ($a = 0.1$, $b = 0.1$, and $c = 3.0$) these equations have a periodic attractor that loops through a three dimensional phase space, as shown in Fig. 8.8. In this chapter we will explore the behavior of the Rössler equations for a different set of parameter values: $a = 0.1$, $b = 0.1$, and $c = 10$. We will see that the dynamics of the Rössler system for these parameter values are quite different from the periodic dynamics we saw in the previous chapters.[1]

---

1. It is traditional to use a different set of equations, known as the Lorenz equations, to introduce chaos in differential equations. These equations, introduced by Edward Lorenz in 1963, played an important role in the advancement

9.1 Chaos in Three Dimensions

In Fig. 9.1 I have plotted the solutions to the Rössler equations
with the parameters given in the caption. For initial conditions,
I chose $x(0) = 5$, $y(0) = 5$, and $z(0) = 1$. The solution consists of
three functions: $x(t)$, $y(t)$ and $z(t)$, shown separately in Fig. 9.1.
Let's take a look. The x and y variables wiggle at a fixed frequency.
But if you look closely, you'll see that the amplitudes of the wiggles
are not constant; some peaks are higher than others. That said,
looking at just $x(t)$ and $y(t)$, one wonders if the behavior might
be periodic.

But take a look at at $z(t)$, shown in the bottom plot of Fig. 9.1.
Here we see upward spikes at regular time intervals. The heights
of these spikes, however, are most certainly not regular: the spikes
have different heights, and there does not appear to be a pattern.
As you might have anticipated, the trajectories for the solutions are
aperiodic: they do not repeat. You are probably wondering what
the solutions of Fig. 9.1 would look like plotted in phase space.
We'll look at such a plot momentarily. First, however, let's think
about aperiodicity, sensitive dependence on initial conditions, and
chaos.

We first encountered aperiodic behavior in Section 4.3 when
we were considering the orbits of the logistic equation, an iterated
one-dimensional function. Here we have found aperiodic behavior

---

of chaos and the discovery of strange attractors. I'm breaking from this tradi-
tion for two reasons. First, the Lorenz system is covered thoroughly in many
other references, including Strogatz (2001, Chapter 9), Peitgen et al. (1992,
Section 12.4), and Hilborn (2002, Section 1.5). Lorenz and the role that his
work played in developing non-linear dynamics are nicely discussed in the popu-
larizations by Gleick (1987) and Stewart (2002) and the history of dynamical
systems by Aubin and Dahan Dalmedico (2002). See also Lorenz's memoir
(1993). The second reason that I am not using the Lorenz equations to illustrate strange attractors is because I think that the Rössler attractor's dynamics are
a bit simpler to visualize and provide a particularly clear view of how the attractor stretches and folds trajectories—the geometric action that is at the heart of
chaos.

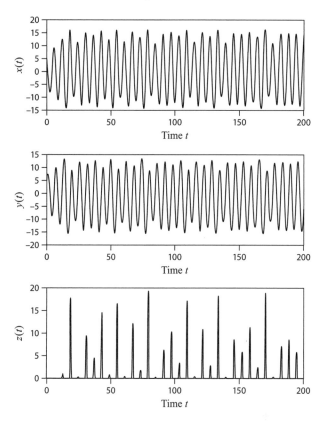

Figure 9.1. Solutions $x(t)$, $y(t)$, and $z(t)$ to the Rössler system, Eq. (9.1). The parameter values are $a = 0.1$, $b = 0.1$, and $c = 10.0$. The initial conditions are $x(0) = 5$, $y(0) = 5$, and $z(0) = 1$.

in the Rössler equations, a three-dimensional differential equation. Recall that for a dynamical system to be considered chaotic, it must have the following properties:

1. Its time evolution is given by a deterministic function.
2. Its orbits are bounded.
3. Its orbits have sensitive dependence on initial conditions.
4. Its orbits are aperiodic.

I stated this definition in Section 4.6 in the context of iterated functions, but it applies to differential equations, too. The Rössler system is a deterministic dynamical system that has bounded, aperiodic orbits. Does it have sensitive dependence on initial conditions? Let's see.

To test for sensitive dependence on initial conditions—that is, the butterfly effect—I'll do what I did in Section 4.4 and plot the solutions to the differential equation for two slightly different sets of initial conditions. The results of doing this are shown in Fig. 9.2. The solid curve is the solution for the initial condition $x(0) = 5$, $y(0) = 5$, and $z(0) = 1$. The dotted line is the solution for a slightly different initial condition: $x(0) = 5.001$, $y(0) = 5$, and $z(0) = 1$.

As expected, the solutions for the two different initial conditions start off essentially identical. It isn't until around $t = 150$ that the solutions become significantly different and the dashed curve becomes visible. This is the butterfly effect: sensitive dependence on initial conditions. Two solutions with almost identical initial conditions eventually become very far apart. Note that the initial conditions are different only for the x variable. The other two variables are initially identical—they both have $y(0) = 5$ and $z(0) = 1$. Nevertheless, the y and z solutions are not identical. This is to be expected, since we are working with a set of three coupled differential equations. Equation 9.1 says that the rates of change of each variable depend on the others; their fates are intertwined.

I've discussed the butterfly effect in some detail in Chapter 4 and 5; I won't repeat that discussion here. But there are two things I want to highlight before moving on. First, recall that one of the phenomena we encountered when examining sensitive dependence on initial conditions for the logistic equation was that greatly increasing the accuracy of our initial condition results in only a small improvement on the accuracy of our solutions. Suppose that in Fig. 9.2, the solid curve represents the system's true

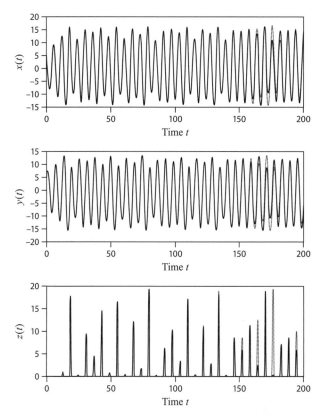

Figure 9.2. Solutions $x(t)$, $y(t)$, and $z(t)$ to the Rössler system, Eq. (9.1). The parameter values are $a = 0.1$, $b = 0.1$, and $c = 10.0$. The initial condition for the solid curves is $x(0) = 5$, $y(0) = 5$, and $z(0) = 1$. For the dotted curve the initial condition is $x(0) = 5.001$, $y(0) = 5$ and $z(0) = 1$.

behavior and the dotted curve is our prediction of the system's behavior. Our measurement of the x initial condition is off by 0.001, or 0.02%. With this degree of accuracy we are able to predict reliably out until around $t = 150$. What if we measure 1000 times more accurately, so that our initial condition is $x(0) = 5.000001$? In Fig. 9.3 I have shown the $z(t)$ solutions for two

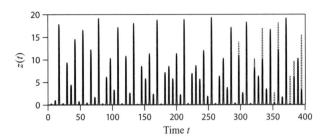

Figure 9.3. The z part of the solutions to the Rössler system, Eq. (9.1). The parameter values are $a = 0.1$, $b = 0.1$, and $c = 10.0$. The initial condition for the solid curves is $x(0) = 5$, $y(0) = 5$, and $z(0) = 1$. For the dotted curve the initial condition is $x(0) = 5.000001$, $y(0) = 5$ and $z(0) = 1$.

initial conditions that are identical, except for one $x(0) = 5.0$, and for the other, $x(0) = 5.000001$. The two trajectories start to separate around $t = 300$. So our measurement accuracy has increased by 1000 times, but we are only able to make accurate predictions for twice as long. This is the butterfly effect again.

The second thing I want to highlight is the irregular nature of the solutions to the Rössler equations. The $z(t)$ solutions, as seen in the bottom plot of Fig. 9.1 are, well, a bit weird. The spikes occur at regular intervals, but the spikes themselves seem highly irregular. As we saw with the logistic equation in Chapters 4 and 5, we are are again seeing a deterministic system produce apparently random behavior. Imagine the situation reversed—suppose we had obtained the $z(t)$ curve from some experimental data: perhaps the strengths of spiking neuronal signals or the amount of rain in yearly monsoon seasons. One might be led to model or understand the situation using probability or a stochastic model of some sort, not realizing that a simple, deterministic system such as Eq. (9.1) could produce such jittery output. We will return to the $z(t)$ spikes in Section 9.4, where we will see that they can be explained by a one-dimensional iterated function.

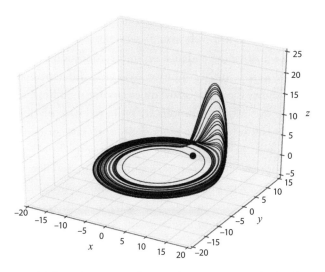

Figure 9.4. The solutions to the Rössler system, Eq. (9.1), plotted in phase space. The parameter values and initial conditions are the same as those of Fig. 9.1. The initial condition is indicated with a black circle.

9.2 The Rössler Attractor

Now it's on to the main attraction of this chapter: what happens if we plot the three solutions, $x(t)$, $y(t)$, and $z(t)$ together in phase space? These solutions are aperiodic: they don't repeat. So perhaps the trajectory in phase space would look like a tangled ball of yarn, a mess of trajectories weaving in and out, filling up a three-dimensional region of phase space. Let's take a look. The solutions from Fig. 9.1 are plotted in phase space in Fig. 9.4. The shape is definitely not a messy ball of yarn. The motion in the x–y plane is mostly circular, as one might expect from looking at the $x(t)$ and $y(t)$ solutions in Fig. 9.1. The circular motion is counter-clockwise when viewed from above. The spikes in the $z(t)$ solution on the bottom of Fig. 9.1 are what lead to the "handle" on the upper right of the phase space plot in Fig. 9.4.

This is an intriguing shape. Would it look different if I chose a different initial condition? It would seem that it must. The Rössler system has sensitive dependence on initial conditions, and so we know that different initial conditions will yield different solutions. Figure 9.5 shows the phase space trajectory for six different initial conditions. The initial conditions are indicated with black circles. I've left the axes off of these plots in order to save space. We see that in all cases the trajectory, perhaps after some brief transient behavior, approaches a shape very similar to that which we saw in Fig. 9.4. How can this be? This seems contrary to the butterfly effect that we observed in Figs. 9.2 and 9.3. In those figures we saw that two initial conditions that differ only slightly give rise to very different trajectories.

The shape shown in phase space in Figs. 9.4 and 9.5 is an example of a *strange attractor*. It is usually referred to as the Rössler attractor, after Otto Rössler, who put forth Eq. 9.1 in 1976 (Rössler, 1976). The Rössler attractor is a fairly complex shape in phase space. It's a band that stretches up and then folds on itself. Although the phase space is three dimensional, the attractor itself is essentially two dimensional.[2] I'll say more about the stretching and folding of the attractor in Section 9.5. But first, let's dig into strange attractors. What is strange about them, and in what sense are they attractors?

In Fig. 9.5 we observed that six different initial conditions all yield a qualitatively similar shape in phase space. The shape is thus an attractor, pulling nearby trajectories into it. Here is another way to see that the Rössler attractor is indeed attracting. In the

---

2. Actually, the dimension of the Rössler attractor is just a little bit larger than two (Peitgen et al., 1992, p. 644): between 2.01 and 2.02. The notion of dimension used here is the box-counting or capacity dimension (sometimes also called the fractal dimension), which is different from the more familiar topological dimension. The box-counting dimension and related constructions are standard topics in many textbooks on dynamical systems. The *Very Short Introduction* on Fractals by Falconer is a clear, not-very-technical introduction to fractals and dimensions (Falconer, 2013). See also Chapters 16 and 18 of Feldman (2012).

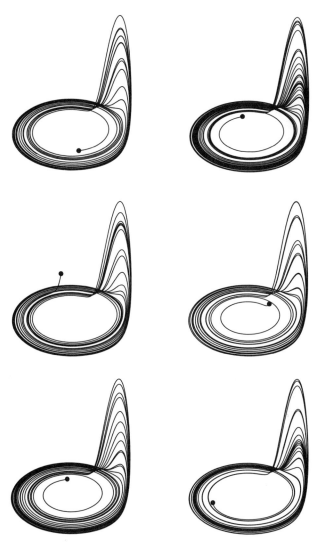

Figure 9.5. Solutions $x(t)$, $y(t)$, and $z(t)$ to the Rössler system plotted in phase space for six different initial conditions. The initial conditions are shown as a solid black circle.

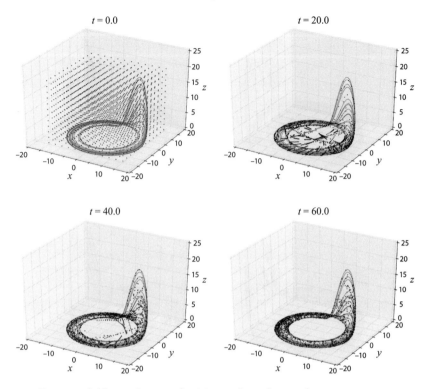

Figure 9.6. The evolution of 1728 initial conditions that start out uniformly spaced on a cube whose x and y values range from -15 to 15 and whose z values range from 0 to 20. The positions evolve according to the Rössler equations. The values of the 1728 initial conditions are shown at $t = 20$, $t = 40$, and $t = 60$.

upper-left plot of Fig. 9.6 I have plotted 1728 (12^3) points evenly distributed on a cube extending from -15 to 15 in the x and y directions and from 0 to 20 in the z direction. I then used these 1728 points as initial conditions for the Rössler equation, Eq. (9.1). I have also plotted the Rössler attractor with a gray line. Where do these 1728 initial conditions end up? By the time $t = 20$, shown in the upper-right of Fig. 9.6, almost all the points are on the attractor or in the middle of the empty circle on the x-y plane inside the attractor. When $t = 40$, most of the points have left the empty inner circle and are on the attractor. And when $t = 60$, all the points are on or very close to the attractor. Figure 9.6 thus shows that the Rössler attractor is indeed an attractor; it pulls in nearby orbits.

The behavior on the attractor itself, however, is chaotic. We saw this back in Fig. 9.2 where I plotted the solutions to the Rössler equations for two almost-identical initial conditions. We observed that the two solutions become quite different around $t = 150$. Note that in Fig. 9.2 the difference between the two $z(t)$ solutions is particularly dramatic. How does the butterfly effect manifest itself in phase space? In Fig. 9.7 I have shown the time evolution of 100 points that start on the attractor. The points are initially uniformly spaced in a square that sits on the flat part of the attractor. The points move together counter-clockwise around the attractor and when they reach the handle of the attractor at $t = 2.5$ they spread out at bit in the z direction. This is shown in the top right plot on Fig. 9.7.

At $t = 6.0$ the points continue on their counter-clockwise journey and are lined up on the outer rim of the circular part of the attractor, as seen in the middle-left plot. The points are then stretched in the z direction when they reach the attractor handle, as shown in the middle-right plot in Fig. 9.7. This general pattern continues. The 100 points move together around the attractor and are stretched when they traverse the handle. At $t = 24.5$ in the lower-left plot on the figure, the points have spread out and now extend along almost the entire lower band of the attractor. At $t = 26.7$ in the lower-right, we again see that the points are spread out as they move through the handle.

Figure 9.7 shows that the trajectories on the attractor are chaotic, but not dramatically so. You might have expected the initial conditions to spread out and uniformly fill the attractor. However, this is clearly not the case. The system definitely has sensitive dependence on initial conditions: initial conditions that

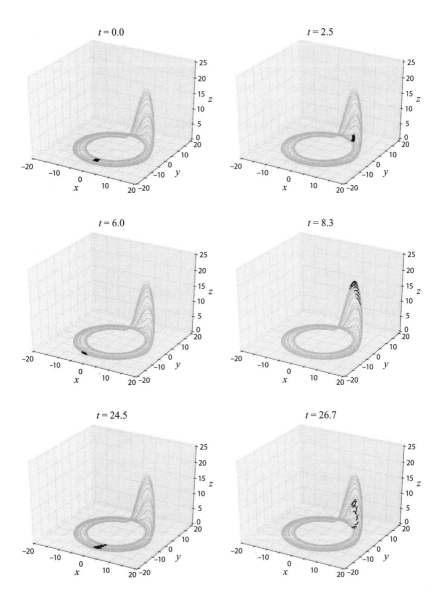

Figure 9.7. The evolution of 100 initial conditions that start out uniformly spaced on a square that sits on the attractor.

start off close to each other have very different z values when they traverse the handle, and the orbits spread out radially on the attractor. However, there is also some regularity to the orbits: the initial conditions move counter-clockwise around the attractor and remain close together.

One can think of the dynamics of the Rössler attractor as consisting of three components. First, the dynamics pull orbits very quickly to the attractor; this is what we observed in Fig. 9.6. Second, orbits move together around the attractor in a counter-clockwise direction. And third, nearby orbits tend to get pulled apart when they move through the handle in the attractor. This last aspect is what is responsible for sensitive dependence on initial conditions.[3] Thus, it is the stretching in the z direction that is responsible for the chaotic behavior in the Rössler equations. This suggests that perhaps the behavior of the Rössler orbits are effectively one-dimensional. We shall see in Section 9.4 that this is indeed the case.

9.3 Strange Attractors

Before moving on, I want to talk about strange attractors in more general terms. Recall that we saw in Section 8.4 that bounded, aperiodic orbits were not possible for two-dimensional systems of differential equations. Trajectories of a differential equation cannot cross—doing so would violate determinism—and so in two

---

3. This qualitative analysis of the dynamics of the Rössler system can be made quantitative by calculating the Lyapunov exponents (see Section 4.7). Since the Rössler system is three-dimensional, there are three Lyapunov exponents. Numerical estimates for their values are 0.0714, 0, and −5.3843 (Sprott, 2003, p. 431). Recall that a positive exponent is associated with sensitive dependence on initial conditions. For the Rössler system there is one positive exponent, 0.0714. Note that while positive, this is fairly small. In contrast, the Lyapunov exponent for the logistic equation with $r = 4$ is $\ln 2 \approx 0.693$. So orbits are pushed apart in the Rössler system, but not as quickly as they are for the logistic equation. The zero Lyapunov exponent corresponds to the circular motion along the attractor, while the negative exponent of −5.3843 corresponds to the strong attraction to the attractor itself.

dimensions as a trajectory proceeds in phase space it creates barriers that it cannot subsequently traverse. This is not the case in three dimensions, and thus in three or more dimensions, bounded aperiodic behavior—and thus strange attractors—are possible. Strange attractors are fairly common; they are most certainly not a quirk of the Rössler system. They are found in many other systems of three or more differential equations. Strange attractors are also often observed in discrete dynamical systems in two or more dimensions—that is, in iterated functions of two or more variables.[4]

Strange attractors are stable, in the sense that almost all initial conditions get pulled to the strange attractor. Any initial condition we choose will quickly approach the same shape in phase space, as we saw in Figs. 9.5 and 9.6. But the dynamics on the attractor are chaotic. Trajectories have sensitive dependence on initial conditions, and thus long-term prediction is not possible. Strange attractors thus combine order and disorder. They are globally stable, or ordered, in that we know that any trajectory will trace out the strange attractor in the long run. Two different initial conditions will lead to the emergence of the same strange attractor. But dynamical systems with strange attractors are locally unstable. The particular trajectories of two nearby initial conditions will be very different, even as they trace out the same shape in the long run.

A familiar example of local instability paired with global stability is weather and climate. The weather is unpredictable. Weather forecasts in most locations beyond a few days are notoriously unreliable, and weather predictions more than a week in advance are essentially worthless. Forecasts a week or more in the future may be meaningful in locations such as a desert where the weather almost never changes. I live in the northeastern US, where the weather is

---

4. A classic example is the Hénon attractor, introduced by Michel Hénon in 1976. This is a two-dimensional, discrete dynamical system: $x_{n+1} = y + 1 - ax^2$, $y_{n+1} = bx_n$. The Hénon system is covered in many dynamical systems texts. A particularly clear and thorough treatment is Peitgen et al. (1992, Section 12.1). See also Feldman (2012, Chapter 26).

highly variable in all seasons. Here, weather predictions even five or six days ahead are highly suspect. Nevertheless, there is some long-term regularity and stability to the weather over periods of hundreds of years. Temperature and precipitation vary seasonally but follow some regular patterns. For example, where I live it is almost never hotter than 90° F in July and almost never warmer than 40°F in December. In this analogy, the weather is a particular trajectory on the strange attractor, and the attractor itself is the climate. Particularly obtuse climate deniers sometimes like to assert that it is impossible to know if the climate is changing over a period of decades, because we can't predict the weather for more than a few days. This is codswallop. There is absolutely nothing contradictory about there being long-term statistical predictability in a system whose trajectories are unpredictable due to the butterfly effect.

The term "strange attractor" was first used by David Ruelle and Floris Takens in 1971 (Ruelle and Takens, 1971). Although they don't state why they chose this term, I think there are two reasons why one might call strange attractors "strange." First, strange attractors have an irregular, fractal shape. This is very different from the simple shapes of attracting points and cycles. Second, the dynamics on a strange attractor are chaotic. That chaotic behavior can be stable was, I believe, not assumed to be possible before 1970. But we now know that stable chaotic behavior is not just possible, but is quite common in dynamical systems. At the risk of relying on a cliché, strange attractors show an example of order within chaos. We see behavior that is unpredictable due to the butterfly effect, but is also structured and constrained.

9.4 Back to 1D: The Lorenz Map

We now turn our attention back to the Rössler attractor shown in Fig. 9.4. Let's imagine following a single trajectory around the attractor. The journey is generally predictable and unsurprising; it moves counter-clockwise around the attractor. The only

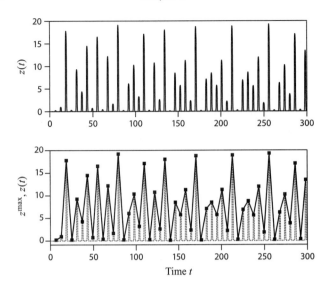

Figure 9.8. The top plot shows $z(t)$ solution to the Rössler system, Eq. (9.1). The bottom plot shows $z(t)$ in gray, while the local maxima for $z(t)$ are plotted with squares.

unpredictable part is when it traverses the upper-left handle. Sometimes the trajectory rises quite high, other time it barely leaves the horizontal plane on which most of the attractor lies. This can be seen in the top plot of Fig. 9.8, where I have plotted the $z(t)$ solution for the Rössler system. As expected, we see that z is usually near zero; this corresponds to the portion of the trajectory that is on the flat, circular portion of the attractor. And $z(t)$ spikes up at regular intervals as the trajectory climbs the attractor's handle.

Let's look at the relative maximum values of z: the heights of the peaks in Fig. 9.8. I'll denote these values as z_1^{max}, z_2^{max}, z_3^{max}, and so on. If you stare at Fig. 9.8 for a bit, you will notice a pattern: a large z^{max} is always followed by a very small one. And a small z^{max} is usually followed by a much larger one. To help us see this pattern more clearly, in the bottom part of Fig. 9.8 I have plotted the local z maxima as squares. The continuous trajectory $z(t)$ is included on the plot in gray.

Note that what we have done is convert the continuous function $z(t)$ into a time series: a sequence of values at a series of discrete time values. For the plot shown in Fig. 9.8, the first ten time series values happen to be:

$$0.08, 0.97, 17.74, 0.29, 9.38, 4.47, 14.56, 0.77, 16.57, 0.41 \,.$$
$$(9.2)$$

The time series plot in the bottom of Fig. 9.8 looks a lot like the chaotic time series generated by the logistic equation back in Chapter 4. See, for example, Fig. 4.7. Recall that in Chapter 4 time series were generated from the logistic equation with the parameter $r = 4$: $f(x) = 4x(1-x)$. We started with an initial condition x_0 and determined x_1, the next value in the time series, by applying the function. That is, $x_1 = f(x_0)$. To get the next iterate, we apply f again; $x_2 = f(x_1)$. And, in general,

$$x_{n+1} = f(x_n) \,.$$
$$(9.3)$$

When working with the logistic equation, as we did in Chapter 4, we know the function $f(x)$, and we use it to figure out x_{n+1} given our knowledge of x_n. What we're going to do now is do this process the other way around. We'll use our knowledge of $z_n^{\max}$ and $z_{n+1}^{\max}$ to figure out the function $f(z^{\max})$ that is generating the time series. To do so, one forms pairs of $z_{n+1}^{\max}$, $z_n^{\max}$ pairs, and plots them. An example will make this clearer. The first ten $z^{\max}$ values are listed in Eq. (9.2). Since the idea is that the function f is generating subsequent values of $z^{\max}$, we know that $f(0.08) = 0.97, f(0.97) = 17.74, f(17.74) = 0.29$, and so on. We can then use this information to form a plot of $f(z^{\max})$. The results of doing so are shown in Fig. 9.9. We see that the points form a nice curve, suggesting that $z_{n+1}^{\max}$ really is a function of $z_n^{\max}$. The construction that leads to Fig. 9.9 is known as the *Lorenz map*, having been introduced by Edward Lorenz (1963) around a half

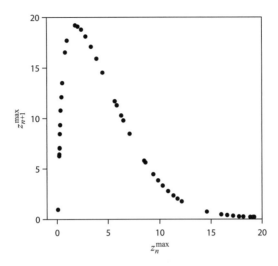

Figure 9.9. The Lorenz map for the Rössler system. This figure was made using the time series shown in the bottom plot of Fig. 9.8.

century ago. The Lorenz map is now a fairly standard tool in the analysis of strange attractors: for example see, Peitgen et al. (1992, Sections 12.4 and 12.5) or Strogatz (2001, Section 9.4).

Let's step back for a moment. The plot shown in Fig. 9.9 is rather remarkable. Although I asserted that we would expect to see that a function was generating the time series of z^{max} values shown in the bottom portion of Fig. 9.8, it is not at all obvious that this is how things were going to work out. It could have been the case that when plotting z_{n+1}^{max} versus z^{max} we would have ended up with a scattering of points suggesting that there was not a functional relationship between the two. But instead we see a very clear relationship; each input has one and only one output.[5] So what Fig. 9.9 demonstrates is that the dynamics of the

5. Actually, this isn't quite true. If we plotted very many points, we could begin to notice a bit of thickness to the line, indicating that z_{n+1}^{max} is not quite completely determined by z_n^{max}.

Rössler equations—a system of three differential equations—can be captured by a one-dimensional iterated function. This realization helps explain a lingering mystery from Chapter 7. There, we saw that certain features of the period-doubling route to chaos in one-dimensional iterated functions are universal. We also saw that the same universal features are observed in higher-dimensional physical systems that seem to be completely unrelated to one-dimensional functions. The Rössler system is not a physical system, but it is three-dimensional and continuous. But we've just seen that its dynamics are captured by a one-dimensional, iterated function. The Rössler system is not anomalous; many other higher-dimensional systems can be reduced to one-dimensional maps. This suggests that we should not be surprised to see universal behavior in higher-dimensional systems, just as we have with one-dimensional maps.

9.5 Stretching and Folding

There is also a geometric relationship between one-dimensional maps and higher-dimensional systems. Any chaotic system must be composed of two elements: stretching and folding. Nearby orbits are stretched apart; this is what gives rise to sensitive dependence on initial conditions. But if a dynamical system only stretched, its orbits would not be bounded. So a chaotic system must involve folding to bring orbits back together.

It is not hard to see that trajectories are stretched apart in the Rössler attractor. This occurs when the trajectories move through the protruding handle that juts up in the z direction. We can see this in Fig. 9.4, where we see the trajectory lines pull apart from each other. As trajectories complete their journey through the handle, they are folded on top of each other. A top view of the attractor, shown in Fig. 9.10, makes this clearer. Here we are looking straight down on the attractor shown in "three dimensions" in Fig. 9.4. A fold is evident in the top right portion of Fig. 9.10.

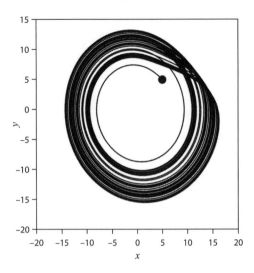

Figure 9.10. A top view of the Rösser attractor. The same attractor is shown in "3D" in Fig. 9.4.

There one can see that trajectories that are roughly in the outer third of the loop are folded over and end up in the inner portion of the attractor.

Stretching and folding also occurs in the logistic equation. This is illustrated in Fig. 9.11, in which I have plotted the logistic equation for $r = 4$: $f(x) = 4x(1-x)$. We saw in Chapter 4 that iterating this equation yielded chaotic orbits. In Fig. 9.11 I have shown what happens to three different initial conditions when iterated once. The initial conditions are $x_0 = 0.05$, $x_0 = 0.1$, and $x_0 = 0.8$, indicated on the figure with a square, circle, and triangle, respectively. The values of these initial conditions after one iteration are shown on the vertical axis.

After the function is applied to them, the square and the circle move away from each other. The square and the circle are farther apart after f is applied to them. So these two points are stretched apart. This will be the case for any two nearby points

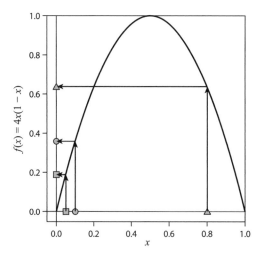

Figure 9.11. Stretching and folding in the logistic equation, $f(x) = 4x(1-x)$.

in a region where the function f has a derivative greater in magnitude than 1 (i.e., $|f'(x)| > 1$). In contrast, the circle and the triangle end up closer together after the function is applied to them. This is folding. So the logistic equation stretches and folds trajectories in a similar way to what happens to trajectories on the Rössler attractor.

As noted previously, stretching and folding are the essential geometric ingredients for chaos. Stretching is needed to give rise to sensitive dependence on initial conditions, while folding keeps orbits bounded. One-dimensional maps like the logistic equation stretch and fold, and thus capture some features of higher-dimensional chaotic systems. We have seen this for the Rössler equation, where the dynamics can be captured by the one-dimensional Lorenz map, shown in Fig. 9.9. The Rössler equation also exhibits a period-doubling route chaos that is universal. The transition is characterized by $\delta \approx 4.669$, just as we saw for the logistic equation (see, e.g., Strogatz (2001, pp.376–9)).

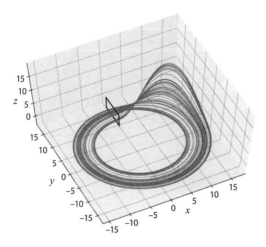

Figure 9.12. The Rössler attractor and a Poincaré section. The Poincaré section is a rectangle in the $x = 0$ plane.

9.6 Poincaré Maps

In constructing the Lorenz map in Section 9.4 we took a three-dimensional differential equation whose solutions are three continuous functions $(x(t), y(t), z(t))$ and reduced it to an iterated one-dimensional function, shown in Fig. 9.9. Iterating the function yields a time series of z^{max} values. So the Lorenz map took a three-dimensional differential equation and turned it into a one-dimensional iterated function. This is quite a simplification. A construction closely related to the Lorenz map is the Poincaré map. Like the Lorenz map, the Poincaré map has the effect of lowering the dimension of the dynamical system under study. In this section I'll describe the Poincaré map.

Consider a continuous trajectory in phase space, such as the three-dimensional Rössler equation trajectory shown in Fig. 9.4 and re-drawn in Fig. 9.12. To understand the attractor's structure and dynamics, it can be useful to examine a cross section. To do so, one chooses a lower-dimensional surface, referred to as a *Poincaré*

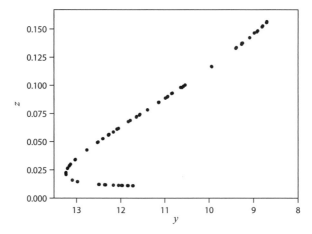

Figure 9.13. A Poincaré section for the Rössler attractor. The Poincaré section is shown in Fig. 9.12.

section and looks at where the attractor intersects that surface. I chose for my Poincaré section a rectangle[6] in the $x = 0$ plane. The Poincaré section—that is, the rectangle— is shown in Fig. 9.12 along with the Rössler attractor. All the trajectories flow through the rectangle. It appears as if the trajectories flow underneath, but that is an illusion caused by the necessity of visualizing a three-dimensional figure on a two-dimensional sheet of paper. Note that in Fig. 9.12 we are looking at the Rössler attractor at a slightly different perspective than in the previous figures.

The Poincaré section is shown in Fig. 9.13. Each intersection of the attractor with the rectangle is shown as a small circle. The Poincaré section was chosen to occur shortly after the trajectories complete their traversal of the handle that juts upward. This handle carries out a fold; the Poincaré section shows the fold as it is almost complete. The Poincaré section for the Rössler attractor is not super exciting. For other systems, the Poincaré section

6. The rectangle's four corners are $(0, 9.5, -3)$, $(0, 9.5, 1)$, $(0, 16.5, -3)$, and $(0, 16.5, 1)$.

often shows an interesting fractal structure. Poincaré sections are not just tools for visualization. They can also be used to analyze dynamics. Let $\vec{x}_0$ be the first point[7] on the Poincaré section— that is the first time the trajectory intersects the rectangle. Denote by $\vec{x}_1$ the next point on the Poincaré section, and so on. Since the dynamical system is deterministic, the location of one point on the Poincaré section determines the location of of the next. There thus exists a function that gives the value of the next point on the Poincaré section given the previous one. This function, or map, is called the *Poincaré map*. Calling this map P, we then have $P(\vec{x}_n) = \vec{x}_{n+1}$. When analyzing the behavior of a dynamical system in some cases it is easier to work with the Poincaré map P than the original dynamical system. Poincaré maps are a standard topic in most intermediate to advanced dynamical systems textbooks. (See e.g., Strogatz (2001, Section 8.7) or Hilborn (2002, Section 5.2).)

9.7 Delay Coordinates and Phase Space Reconstruction

Let's revisit the Lorenz map from Section 9.4 and think about its construction. We started with $z(t)$ in Fig. 8.7, and we then determined the values of the local maxima and used these values to produce the Lorenz map in Fig. 9.9. Note that in so doing we did not need to use the equations that generated $z(t)$. Simply having $z(t)$ in our possession was enough. In our case $z(t)$ arose from solving the Rössler equations on a computer, but it could have its origin in an experiment in which the equations generating $z(t)$ are unknown.

Constructing the Lorenz map is an example of *non-linear time series analysis*, a broad area of applied mathematics that takes notions and ideas from nonlinear dynamics and applies them to time series with the goal of characterizing their qualities and, often,

---

7. I'm writing this point as $\vec{x}$ and not x, since points on the Poincaré section will typically be two-dimensional coordinates, not scalars.

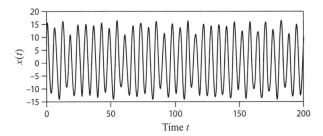

Figure 9.14. A time series $x(t)$.

producing predictive models. *Time series* in this context refers to a long sequence of measurements indexed by time. The measurements could be the result of a physical experiment: the successive positions of a pendulum or a sequence of hourly temperature measurements. Time series also arise from biological and social systems, and can also be generated by a simulation. Nonlinear time series analysis is a large and growing field. I'll conclude this chapter by giving a sketch of a fun and surprising technique from nonlinear time series analysis: using delay coordinates to reconstruct an attractor.

Suppose we have a long series of measurements which I'll denote $x(t)$. An example is shown in Fig. 9.14. This time series happens to be an $x(t)$ solution to the Rössler equation. The technique I'm about to describe, however, can be applied to any time series, even if we don't know its origin. The time series in Fig. 9.14 looks erratic. We might wonder if it arose from a chaotic dynamical system. Could $x(t)$ be part of a higher-dimensional dynamical system? That is, perhaps there is also a $y(t)$ and $z(t)$ that is part of this system, but we don't know about these variables and/or aren't able to measure them. And might this higher-dimensional system possess a strange attractor? These are good questions that seem unanswerable. All we have at our disposal is $x(t)$—a one-dimensional signal. How can we build a multi-dimensional phase space? Is this the end of the road?

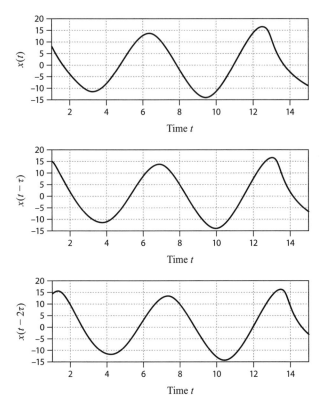

Figure 9.15. The original signal $x(t)$ and the two delay coordinates, $x(t - \tau)$ and $x(t - 2\tau)$.

Amazingly, there is a way forward. We proceed by using a clever construction. We choose a time delay τ and then form a new time series $x(t - \tau)$ which is $x(t)$ shifted backwards in time (i.e., delayed—by τ). In the top two plots of Fig 9.15 I have have plotted $x(t)$ and $x(t - \tau)$. To make these plots I chose $\tau = 0.5$. (Shortly I will give a not-very-satisfactory reason for this choice.) Note on the graph that $x(t - \tau)$ is $x(t)$ shifted to the right by τ units. The function $x(t - \tau)$ is known as a *delay coordinate*.

There is no reason to limit ourselves to just one delay coordinate. We can form another delay coordinate, $x(t - 2\tau)$. As

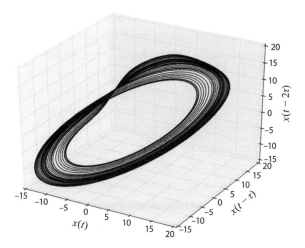

Figure 9.16. The Rössler attractor "reconstructed" using delay coordinates with $\tau = 0.5$.

shown in the bottom plot in Fig. 9.15, $x(t - 2\tau)$ is $x(t)$ shifted to the right by 2τ time units. As before, I've chosen $\tau = 0.5$. The three functions $x(t)$, $x(t - \tau)$ and $x(t - 2\tau)$ extend from $t = 1$ to $t = 200$. However, in Fig. 9.15 I have shown them from 1 to 15 to allow you to see how the peaks in delay coordinates are shifted to the right. The shift is perhaps most easily seen if you note the positions of the right-most peak in the three plots.

We now have three functions: $x(t)$, $x(t - \tau)$, and $x(t - 2\tau)$. This state of affairs is similar to Fig. 9.1, where our three functions were the solutions to the Rössler equations: $x(t)$, $y(t)$, and $z(t)$. This suggests that we plot $x(t)$, $x(t - \tau)$, and $x(t - 2\tau)$ in "phase space" and see what results. I put phase space in quotes, because this is sort of a phony phase space. Via delay coordinates we've bootstrapped our way to a three-dimensional phase space even though we really just have a one-dimensional signal, $x(t)$. Nevertheless, let's plot the delay coordinates in a three-dimensional space and see what happens. The results of doing this are shown in Fig. 9.16.

Kinda weird, eh? Figure 9.16 looks an awful lot like the Rössler attractor in Fig. 9.4. The two figures aren't identical, but are very similar: both consist of a circular band, a portion of which folds over on itself. The Rössler attractor, Fig. 9.4, arose from plotting the three solutions $x(t)$, $y(t)$, and $z(t)$ against each other. But in Fig. 9.16 we recovered a near facsimile of the Rössler attractor using *only* $x(t)$. How can this be?

The Rössler equations are coupled. In other words, $x(t)$ depends on $y(t)$ and $z(t)$. So the single time series $x(t)$ contains information about $y(t)$ and $z(t)$ as well. Thus it is perhaps not that surprising that we can produce a shape like the Rössler attractor using $x(t)$ alone, since at some level $x(t)$ "knows" about $y(t)$ and $z(t)$. More formally, one can show that under appropriate conditions, the mapping from delay coordinates $(x(t), x(t-\tau), x(t-2\tau))$ to the original coordinates $(x(t), y(t), z(t))$ is one-to-one. The original (x, y, z) dynamics are deterministic, and so because of the one-to-one relation between $(x(t), x(t-\tau), x(t-2\tau))$ and $(x(t), y(t), z(t))$, the dynamics of $(x(t), x(t-\tau), x(t-2\tau))$ are deterministic as well. So this says that the dynamics in our phony phase space $(x(t), x(t-\tau), x(t-2\tau))$ are the same as the original dynamics. All we've done is changed coordinates from $(x(t), y(t), z(t))$ to $(x(t), x(t-\tau), x(t-2\tau))$. We haven't changed the dynamics. The attractor in the original and phony phase spaces may look different, but they're generated by the same dynamics.

Moreover, generically it will be the case that the coordinate change is not only one-to-one, but is also continuous and smooth. In this case, the attractors in the original and phony phase spaces will have the same properties—the same dimension and the same Lyapunov exponents. And so what I've been calling the phony phase space isn't really phony at all; it's legit. The process of using a single time series and then forming delay coordinates to make a phase space is called *phase space reconstruction*. It is also sometimes referred to as *attractor reconstruction* or *embedding*. The picture behind the latter term is that we are taking an attractor (that might live in a high-dimensional phase space) and "embedding" it in different space.

As you have probably sensed, there is a good bit of mathematical detail that I've swept under the rug. A full discussion of these details would take us too far afield, but there are two big questions that I want to at least say something about. How do we know how many delay coordinates to choose? And how do we choose the time delay τ? Regarding the number of delay coordinates, there is some good news: there is a theorem that provides guidance. The number of delay coordinates must be larger than twice the box-counting dimension of the attractor one is trying to reconstruct. But there's also some bad news: the dimension of the attractor is almost always not known when one applies phase space reconstruction: the starting point is a time series $x(t)$ of unknown origin. So this theorem doesn't help much in practice. This story ends with some good news: in practice, usually figuring out the number of coordinates needed, known as the *embedding dimension*, is not difficult. One increases the number of delay coordinates and looks to see when an attractor or limit cycle emerges.[8]

The issue of how to choose the time delay τ is more difficult. One seeks a τ that is neither too small nor too large. Suppose we use a super-small τ: say $\tau = 0.001$. Then $x(t)$ and $x(t - 0.001)$ are almost the same; $x(t - 0.001)$ tells us almost nothing that we didn't already know from $x(t)$. We want to choose a τ that is long enough so that some of the information from $y(t)$ and $z(t)$

---

8. Of course one wouldn't know in advance what the attractor should look like. So in practice determining the embedding dimension is not done by eye alone. One approach is to look at neighboring points in the embedding space. These neighboring points should move together, at least for short time ranges, since we assume that the dynamical system yields continuous solutions. If neighboring points do not move together, this is an indication that the embedding dimension is not high enough to fully unfold the attractor. See e.g., Kantz and Schreiber (2004, Section 3.3).

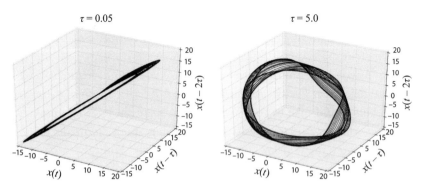

Figure 9.17. The Rössler attractor reconstructed using bad τ values. On the left figure, $\tau = 0.05$. On the right, $\tau = 5.0$.

(or however many dimensions there are in the true phase space) have time to "flow" into $x(t)$. This, after all, is how phase space reconstruction works: by sampling $x(t)$ at different times, we are incorporating information about $y(t)$ and $z(t)$. A reconstructed Rössler attractor with a too-small τ of 0.05 is shown in the left plot of Fig 9.17. Since the three coordinates, $x(t)$, $x(t - \tau)$ and $x(t - 2\tau)$ are almost identical, the plot is an almost-straight line in three dimensions.

On the other hand, suppose we choose a very large τ: perhaps $\tau = 250$. Then $x(t)$ and $x(t - 250)$ will be very different—almost uncorrelated. The problem here is that two much information about $y(t)$ and $z(t)$ will have flowed into $x(t)$. The shape in reconstructed phase space will be complicated and folded over on itself. This situation is illustrated in the right plot of Fig. 9.17. The figure looks pretty, but it doesn't look that much like the Rössler attractor.

There is an approach one can take to choosing τ that is a bit less *ad hoc*. The idea is to look at some measure of how $x(t)$ is correlated with itself, either via the auto-correlation function or the mutual information. One then uses this to choose a time delay τ that is not so small as to be almost perfectly correlated with $x(t)$

but is not so large that the correlation between $x(t)$ and $x(t-\tau)$ is very small. My sense is that τ-selection methods are a bit of an art as well as a science. If you want to pursue this issue further, suggestions are found in the Further Reading section at the end of this chapter.

9.8 Determinism vs. Noise

Phase space reconstruction is a powerful tool. Upon first encounter, it seems almost like magic—that's certainly how it appeared to me. But there are limits to the magic: phase space reconstruction and related methods like the Lorenz map will be successful only if the dynamical system generating the time series $x(t)$ is deterministic. If there is a rule generating $x(t)$ (along with possibly many other unobserved variables), then delay coordinates can be used to reconstruct phase space and study the phase-space dynamics. If the dynamical system is not deterministic, however, then phase space reconstruction will not be successful. Delay coordinates cannot reveal a dynamic that is not there. I'll illustrate this with a simple example.

Let's return one more time to the logistic equation. In the top left plot of Fig. 9.18 I have shown a time series generated with the logistic equation, $f(x) = 4x(1-x)$. In the upper right plot in the figure I have taken that time series and done something very similar to what we did when we made the Lorenz map for the Rösser equations in Section 9.4: I plotted x_{n+1} versus x_n. I used 100 points to make this plot; I only showed 30 in the time series plot. The result in the upper right part of Fig. 9.18 is the logistic equation—the function we used to generate the time series in the first place. In this case this is not a surprise. But if we were given the time series and we didn't know where it came from, the result in the top right of the figure would be big news, telling us that the apparently random time series was generated by a simple one-dimensional iterated function.

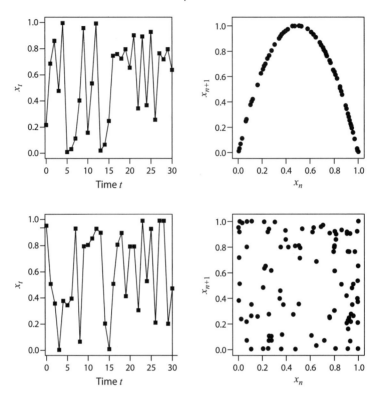

Figure 9.18. The left plots show two time series. The top time series was generated from the logistic equation. The bottom time series was generated stochastically. Iterates were chosen by sampling at random from the distribution generated by the top time series. The right-hand plots show x_{n+1} plotted against x_n.

In the lower left of Fig. 9.18 I have plotted a different time series. This time series was generated by a stochastic process, not a deterministic dynamical system. Recall that a system is stochastic if there is an element of chance in its output; the same input will not always give the same output. To make the time series I determined the distribution that describes the 100 iterates I generated for the top plots, and then drew x values at random from this distribution. I did this by resampling from the original time series, with replacement. A plot of this time series is shown in the lower left part of Fig. 9.18. As expected, it looks different from the logistic time series in the top left of the figure. Nevertheless, if we encountered this time series as the result of some experiment, we might be tempted to wonder if it had been generated by a deterministic dynamical system. The lower right of Fig. 9.18 indicates that this is not the case. The value of x_{n+1} does not appear to be a function of x_n.

There are two take-home messages from Fig. 9.18. First, phase space reconstruction and related methods do not work if the system is not deterministic. Phase space reconstruction is almost magic, but it cannot conjure an attractor out of a stochastic system. Second, the process of phase-space reconstruction gives us a way to determine if a time series is deterministic or stochastic. This is an important thing to know about whatever system you are studying.

In practice, of course, the world is an unknown mix of random and deterministic. How much determinism must there be in a system for non-linear time series analysis to be of use? What level of noise in your measuring device or data-gathering methods will be too much? These are difficult questions that are topics of current investigation.

9.9 Further Reading

Strange attractors are a standard topic that is treated in almost all dynamical systems texts and popular accounts of chaos. An accessible review article on strange attractors is Ruelle (1980). The Rössler system was introduced by Otto Rössler in 1976. For reviews of the Rössler system, see Letellier and Rössler (2006) and Gaspard (2005). A discussion of the motivation and influences on Rössler as he developed his equation, see Letellier and Messager (2010). Chapter VII of Aubin (1998) provides a detailed account of interdisciplinary efforts toward a mathematical understanding of fluid turbulence in the 1970s and 80s, including strange attractors. The illustrated book by Abraham and Shaw (1992) contains many drawings of strange attractors and is a fantastic, visual introduction to dynamics.

Delay coordinates and phase space reconstruction are a subset of the field of nonlinear time series analysis. The standard text on nonlinear time series analysis is Kantz and Schreiber (2004). Chapters 3 and 9 of Kantz and Schreiber (2004) are the clearest introduction to phase space reconstruction that I am aware of. See also Chapter 5 of Sauer (2006). A good, recent review article on nonlinear time series analysis is Bradley and Kantz (2015). A brief introduction to attractor reconstruction is Sauer (2006). A more mathematical review is Takens (2010). See also Pecora et al. (2007). A commonly-used software package that carries out a wide array of nonlinear time series analyses is TISEAN (TIme SEries ANalysis) by Hegger, Kantz, and Schreiber (1999). The idea of delay coordinates was first put forth in Packard et al. (1980) and was subsequently put on a firmer mathematical footing by Takens (1981). Takens' theory was extended a decade later in Sauer et al. (1991).

10

CONCLUSION

In this final chapter I'll give a short summary of what we've covered and then offer some thoughts on some of the broader lessons that nonlinear dynamics might hold for those interested in complex systems.

10.1 Summary

A dynamical system is a mathematical system that changes in time according to a well specified rule. In this book we looked at two types of deterministic dynamical systems: iterated functions and differential equations. In both of these types of dynamical systems we encountered chaos: bounded, aperiodic orbits that have sensitive dependence on initial conditions. Chaos is possible in one-dimensional iterated functions but requires three or more dimensions for systems of differential equations. We investigated sensitive dependence on initial conditions (SDIC), known more famously as the butterfly effect. If a system has SDIC, a tiny, perhaps even imperceptible difference in the initial condition can make a large difference in the dynamical system's orbit. As a practical matter, this means that long-term prediction of chaotic systems is not possible. It calls into question our notion of randomness and its relation to determinism, issues that I explored in Chapter 5.

The character of a dynamical system sometimes changes suddenly when a parameter is varied. These sudden qualitative changes, known as bifurcations, occur in both differential equations and iterated functions. We used bifurcation diagrams to visualize and analyze these transitions. The bifurcation diagram for the logistic equation revealed intricate fractal structure. The logistic equation transitions from periodic to chaotic behavior via a series of period-doubling bifurcations. We saw that features of this transition were universal: the same for almost all one-dimensional functions as well as higher dimensional systems, including differential equations. Universality even extends to physical systems that undergo period-doubling transitions to chaos.

Our study of systems of differential equations led us to strange attractors: attracting structures in phase space on which the dynamics are chaotic. Strange attractors, whose shapes are often elegant and fractal, combine order and disorder. They are locally unstable (individual trajectories are unpredictable due to SDIC), but globally stable (all trajectories in the long run trace out a similar shape). I also touched briefly on phase space reconstruction, a technique that lets one take a single time series and reconstruct the higher-dimensional phase in which it is embedded.

Deterministic dynamical systems generate behavior that is aperiodic and has sensitive dependence on initial conditions. We've seen lots of examples of simple systems producing erratic solutions that seem to be of a totally different character from the regular rule that creates them. And order and disorder live together happily in strange attractors. The study of dynamical systems thus suggests that order and disorder are not opposites or completely separate categories. Ordered rules can produce disordered results.

I think that dynamical systems and chaos are fun and important topics that serve as an excellent entry point into the study of complex systems. For the rest of this chapter I'll offer some personal thoughts about why I think this is so. I'll begin with some remarks about complex systems.

10.2 Complex Systems

I write this section with some trepidation. Complex systems is an evolving area of study and not everyone agrees on what the term complex systems should mean. There is also not full agreement on what term to use; some say "complexity science," while others prefer "complex adaptive systems." I'll stick with complex systems. I don't think all complex systems are adaptive, and I worry that the term "complexity science" suggests that complex systems is a type of inquiry that is epistemically separate or distinct from the rest of science, which I don't think is the case.

There is a spectrum of opinions on the status of complex systems. Is it a discipline, a field, an area of study, a constellation of concepts and methods, an interdisciplinary amalgam of topics, a hodgepodge of ideas, a collection of buzzwords? My own view is toward the middle of this spectrum: I tend to think of complex systems as an area of study that has at its core a semi-solid collection of concepts and methods. I should be clear that "semi-solid" is not meant in a negative way. The boundaries of complex systems are fuzzy and there is not complete agreement on some concepts and definitions. But I think this is a positive state of affairs. I think that other areas of study are sometimes held back by concepts that have become fixed too narrowly. In advocating for fuzzy disciplinary boundaries I'm definitely not advocating for fuzzy thinking. We can recognize the arbitrariness of disciplinary boundaries and the contingent nature of some categories and concepts without sacrificing scientific rigor.

So what is inside the big complex systems tent? Complex systems typically have at least several of the following features.

1. Complex systems consist of many entities that interact. The interactions are strong enough that they cannot be treated as non-interacting.
2. Complex systems are made up of heterogeneous parts.

3. Complex systems are dynamic; they adapt and/or are not in equilibrium.
4. Complex systems often exhibit emergent properties— patterns or structures that are not contained in an obvious way in the description of the system's constituents and their interactions.
5. Complex systems combine elements of order and disorder, and thus resist a simple mathematical analysis.

This list is definitely not meant to be exhaustive or prescriptive. I see complex systems as a notion and not a strict category.

There are methods that are commonly used to study complex systems. Examples include network theory, computation theory, dynamical systems, agent-based models, and simulation more generally. Many of the mathematical tools and techniques used in complex systems have an anti-reductionist flavor. These approaches look for structures and patterns that are common across different systems. Researchers in complex systems draw upon a diverse and interdisciplinary set of mathematical tools and theoretical frameworks. In my view this diversity is a strength of complex systems.

I think that all fields have a duality between the entities they study and the methods they use to study them. For example, physics is the scientific study of the physical world. But it is also a set of methods, techniques, epistemic commitments, habits of mind, and cultural practices. All of these are what make physics physics—a particular way of producing knowledge about some aspects of the world. In order for a research paper to be considered physics, it must be both about the right physics stuff and done the right physics way. The situation is similar with complex systems; it is at once a collection of topics and a collection of mathematical tools and techniques.

In order for a discipline to be a discipline, there needs to be a culture or community of people doing it. There definitely is a community of complex systems scholars. There are conferences, journals, institutes, summer schools, and academic departments dedicated to it. So another approach to thinking about complex systems is anthropological. Find communities that identify as complex systems communities, and what they study and how they study it is what complex systems is. I don't mean to sound glib. I think if you really push on the definitions of other disciplines, you'll often end up in a similar place. Physics is what physicists do. Anthropology is what anthropologists do. And so on. As with any area of study, part of what delineates complex systems as a type of inquiry is a community with shared commitments and aesthetics—a similar eye for scientific or mathematical problems that are considered fun or important.

Complex systems can be thought of as a collection of topics sharing certain features and themes, a collection of methods and techniques, and as a cultural practice. I think dynamical systems is relevant to all these ways of thinking about complex systems.

10.3 Emergence(?)

I discussed the notion of emergence back in Section 7.3 and I'd like to revisit the idea here. An emergent property is generally understood to be a pattern or phenomenon that is not clearly derivable from an understanding of a system's constituent parts and their interactions. I am of two minds regarding emergence. On the one hand, I find it a powerful and compelling idea. When one studies emergent phenomena, one sees patterns that seem to come from nowhere. Simple models produce surprising and delightfully complex results. We've seen examples of this in these pages: the logistic equation's bifurcation diagram and the strange Rössler attractor. Chaotic and periodic behavior in the logistic map can also be seen as emergent. These are features that appear only when the equation is iterated. A deterministic process depending on no historical information—only the present population matters—produces patterns over long time scales.

Emergence is a very appealing idea. Encountering an emergent phenomenon—patterns in nature or a simple simulation that I wrote that produced an unexpected result—is exciting and fun. That said, in my view it has proven difficult to come up with a quantitative, objective definition of emergence. Descriptions of emergence often rely on subjective notions: patterns that cannot *easily* be predicted from knowledge of the constituent parts or structures that are not *obviously* contained in the description of a system's dynamics. Easily or obviously for whom? Pinning down what is meant by emergence has been vexing to philosophers and scientists.

All that being said, appreciation of emergence is one of the important things to come out of the study of dynamical systems, and this appreciation informs the study of complex systems. This certainly is true historically. Many dynamical systems researchers shifted attention to complex systems as the area was coalescing in the 1980s. And scientists in other fields who were drawn in to complex systems were clearly motivated by and influenced by chaos. I think that dynamical systems remains relevant; an awareness and appreciation of the tools and concepts of dynamics is still important for those wishing to learn about and explore complex systems.

Dynamical systems show us very clearly that simple systems do not necessarily possess simple properties. I think this is a fact that is not always appreciated. Many people see a complicated outcome and assume it must have a complicated cause, or see a random outcome and assume it must have a stochastic cause. Cohen and Stewart (1995, pp. 19–22) have a nice phrase that captures this common misunderstanding. They refer to it as the mistaken belief in the "conservation of complexity": the complexity of an explanation should be proportional to the complexity of the phenomenon being explained. Cohen and Stewart give two examples: the eye is a complicated organ, and therefore a simple explanation like evolution by natural selection must be wrong. And inflation is a simple phenomenon and therefore adjusting inflation in the desired direction should be simple (Cohen and Stewart, 1995, p. 29). If we're going to grapple with complex phenomena we need to be clear that complex phenomena do not require complex explanations. Dynamical systems are capable of a range of behavior: simple, complex, predictable, unpredictable. The same can be said of systems of simple interacting entities—this is one of the key lessons of the study of complex systems. So patterns and structures in physical, biological, and social systems are not necessarily evidence of design, intent, or conspiracy. There are many systems that form patterns without any intervention or instruction.

10.4 But Not Everything is Simple

The study of dynamical systems shows us that complex phenomena can have simple explanations. This realization, although perhaps surprising at first, is in many ways not a radical departure from traditional physics, which seeks to explain phenomena in terms of simple, general mathematical laws. Universality is an excellent example of this. A line of attack that began with a simple, abstract model led to the understanding that certain features of the period-doubling route to chaos are universal. Far from overturning traditional physics, universality is perhaps an exemplar of an explanatory mode that is at the heart of physics.

How far can this physics-inspired approach take us? Can simple models elucidate the workings of ever more complex systems? I'm doubtful. Surely simple models will remain an important tool for complex systems. But I don't think it's the case that *all* complex systems have simple explanations. Other styles of explanations will be needed, and many will not be compact and simple. Simply put, complex systems should not be the physics-ization of other fields.

Most scholars working in complex systems share a belief that there are some similar features among many complex systems and that there are common methods and frameworks that can be used to help us find and understand those features. To what extent is this the case? I'm not sure, but I am certain that fun discoveries and debates lie ahead. I would caution against both a reflexive reductionism and a reflexive anti-reductionism. A challenge with complex systems is to remain open to different approaches and to recognize that the relationships between simplicity and complexity are not simple.

10.5 Further Reading

As noted before, much has been written by philosophers and scientists grappling with the notion of emergence. See Bedau and Humphreys (2008), Butterfield (2011*a*), Butterfield (2011*b*), O'Connor and Wong (2015), and references therein. An influential early paper about emergence and complexity is Anderson (1972). See also Crutchfield (2002). Although not about emergence *per se*, Philip Ball's trilogy *Nature's Patterns: A Tapestry in Three Parts* (2011*c*; 2011*b*; 2011*a*) is an engaging and accessible look at pattern formation in a wide range of physical and natural systems. Ball (2016) is a non-technical introduction to pattern formation in nature with breathtaking illustrations.

Much has also been written about complexity and complex systems. For accessible starting points, I recommend Flake (1999), Mitchell (2009), Mitchell (2013), Miller (2016), and Charbonneau (2017). Newman (2011) is an annotated bibliography that gives a sense of the wide scope of complex systems and contains a comprehensive list of references, current as of 2011.

10.6 Farewell

You've reached the end of this short book, but hopefully not the end of your journey into chaos and complex systems. I hope you've found these ideas fun and rewarding to learn about.

BIBLIOGRAPHY

Abraham, Ralph and Shaw, Christopher D. (1992). *Dynamics: The Geometry of Behavior* (2nd ed.). Addison Wesley.

Alligood, Kathleen T., Sauer, Tim D., and Yorke, James A. (1996). *Chaos: An Introduction to Dynamical Systems*. Springer Science & Business Media.

Anderson, Philip. W. (1972). More is different. *Science*, **177** (4097), 393–396.

Arnol'd, Vladimir I. (October 2003). *Catastrophe Theory* (3rd ed.). Springer.

Atmanspacher, Harald and Bishop, Robert C., eds. (2014). *Between Chance and Choice: Interdisciplinary Perspectives on Determinism*. Andrews UK Limited.

Attanasi, Alessandro, Cavagna, Andrea, Del Castello, Lorenzo, Giardina, Irene, Melillo, Stefania, Parisi, Leonardo, Pohl, Oliver, Rossaro, Bruno, Shen, Edward, Silvestri, Edmondo, and others (2014). Finite-size scaling as a way to probe near-criticality in natural swarms. *Physical Review Letters*, **113**(23), 238102.

Aubin, David (1998). *Cultural History of Catastrophes and Chaos: Around the Institut des Hautes Études Scientifiques, France 1958-1980*. Ph. D. thesis, Princeton University. Available at http://webusers.imj-prg.fr/∼david.aubin/publis.html.

Aubin, David (2001). From catastrophe to chaos: The modeling practices of applied topologists. In *Changing Images in Mathematics: From the French Revolution to the New Millenium* (ed. A. Dahan Dalmedico and U. Botazini), pp. 255–279. Routledge.

Aubin, David (2004). Forms of explanations in the catastrophe theory of René Thom: Topology, morphogenesis, and structuralism. In *Growing Explanations: Historical Perspective on Recent Science* (ed. M. N. Wise), pp. 95–130. Duke University Press.

Aubin, David and Dahan Dalmedico, Amy (2002). Writing the history of dynamical systems and chaos: *Longue durée* and revolution, disciplines and cultures. *Historia Mathematica*, **29**(3), 273–339.

Ball, Philip (2011*a*). *Branches: Nature's Patterns: A Tapestry in Three Parts.* Oxford University Press.

Ball, Philip (2011*b*). *Flow: Nature's Patterns: A Tapestry in Three Parts.* Oxford University Press.

Ball, Philip (2011*c*). *Shapes: Nature's Patterns: A Tapestry in Three Parts.* Oxford University Press.

Ball, Philip (2014). One rule of life: Do we all exist at the border of order? *New Scientist*, **222**(2966), 44–47.

Ball, Philip (2016). *Patterns in Nature: Why the Natural World Looks the Way It Does.* University of Chicago Press.

Barnes, Belinda and Fulford, Glenn R. (2002). *Mathematical Modelling with Case Studies: A Differential Equation Approach Using Maple.* Taylor & Francis.

Bedau, Mark A. and Humphreys, Paul E. (2008). *Emergence: Contemporary Readings in Philosophy and Science.* MIT press.

Bishop, Robert C. (2008). What could be worse than the butterfly effect? *Canadian Journal of Philosophy*, **38**(4), 519–547.

Bishop, Robert C. (2017). Chaos. In *The Stanford Encyclopedia of Philosophy* (Spring 2017 ed.) (ed. E. N. Zalta). Metaphysics Research Lab, Stanford University.

Blanchard, Paul, Devaney, Robert L., and Hall, Glen R. (April 2011). *Differential Equations* (4 ed.). Cengage Learning.

Bradley, Elizabeth and Kantz, Holger (2015). Nonlinear time-series analysis revisited. *Chaos: An Interdisciplinary Journal of Nonlinear Science*, **25**(9), 097610.

Britton, Nicholas F. (2005). *Essential Mathematical Biology* (1st ed. 2003. Corrected 2nd printing). Springer.

Butterfield, Jeremy (2011*a*). Emergence, reduction and supervenience: A varied landscape. *Foundations of Physics*, **41**(6), 920–959.

Butterfield, Jeremy (2011*b*). Less is different: Emergence and reduction reconciled. *Foundations of Physics*, **41**(6), 1065–1135.

Chaitin, G. (1966). On the length of programs for computing finite binary sequences. *J. ACM*, **13**.

Charbonneau, Paul (2017). *Natural Complexity: A Modeling Handbook*. Princeton University Press.

Clauset, Aaron, Shalizi, Cosma R., and Newman, M. E. J. (2009). Power-law distributions in empirical data. *SIAM Review*, **51**(4), 661–703.

Cohen, Jack and Stewart, Ian (1995). *The Collapse of Chaos: Discovering Simplicity in a Complex World (Penguin Press Science)*. Penguin (Non-Classics).

Coppersmith, S. N. (1999). A simpler derivation of Feigenbaum's renormalization group equation for the period-doubling bifurcation sequence. *American Journal of Physics*, **67**(1), 52–54.

Coullet, Pierre and Pomeau, Yves (2016). History of chaos from a French perspective. In *The Foundations of Chaos Revisited: From Poincaré to Recent Advancements* (ed. C. Skiadas), Understanding Complex Systems, pp. 91–101. Springer International Publishing.

Coullet, Pierre and Tresser, Charles (1978). Itérations d'endomorphismes et groupe de renormalisation. *Le Journal de Physique Colloques*, **39**(C5), 25–28.

Coullet, Pierre and Tresser, Charles (1980). Critical transition to stochasticity for some dynamical systems. *Journal de Physique Lettres*, **41**(11), 255–258.

Crawford, John D. (1991). Introduction to bifurcation theory. *Rev. Mod. Phys.*, **63**, 991–1037.

Crutchfield, J. P. (2002). What lies between order and chaos. In *Art and Complexity* (ed. J. Casti). Oxford University Press.

Crutchfield, J. P., Farmer, J. D., Packard, N. H., and Shaw, R. S. (December 1986). Chaos. *Scientific American*, **542**(12), 46–57.

Cvitanović, Predrag (1989). *Universality in Chaos* (2nd ed.). CRC Press.

Dakos, Vasilis, Carpenter, Stephen R., Brock, William A., Ellison, Aaron M., Guttal, Vishwesha, Ives, Anthony R., Kefi, Sonia, Livina, Valerie, Seekell, David A., van Nes, Egbert H., and others (2012).

Methods for detecting early warnings of critical transitions in time series illustrated using simulated ecological data. *PloS One*, **7**(7).

DeDeo, Simon (2016). Introduction to Renormalization. A Complexity Explorer Course: https://renorm.complexityexplorer.org/.

Devaney, Robert L. (1989). *An Introduction to Chaotic Dynamical Systems*. Perseus Publishing.

Diacu, Florin and Holmes, Philip (1999). *Celestial Encounters*. Princeton University Press.

Downey, R. G. and Reimann, J. (2007). Algorithmic randomness. *Scholarpedia*, **2**(10), 2574. Revision #90955.

Eagle, Antony (2014). Chance versus randomness. In *The Stanford Encyclopedia of Philosophy* (Spring 2014) (ed. E. N. Zalta).

Edelstein-Keshet, Leah (2005). *Mathematical Models in Biology (Classics in Applied Mathematics)* (1st ed.). SIAM: Society for Industrial and Applied Mathematics.

Ekeland, Ivar (1990). *Mathematics and the Unexpected*. University of Chicago Press.

El Espectador (April 2014). Hipopótamos bravos. *El Espectador*. Editorial.

Ellner, Stephen P. and Guckenheimer, John (March 2006). *Dynamic Models in Biology* (1st ed.). Princeton University Press.

Epstein, Joshua M. (2008). Why Model? *Journal of Artificial Societies and Social Simulation*, **11**(4), 12.

Falconer, Kenneth (December 2013). *Fractals: A Very Short Introduction (Very Short Introductions)* (1st ed.). Oxford University Press.

Feigenbaum, Mitchell J. (1978). Quantitative universality for a class of nonlinear transformations. *Journal of Statistical Physics*, **19**(1), 25–52.

Feigenbaum, Mitchell J. (1983). Universal behavior in nonlinear systems. *Physica D: Nonlinear Phenomena*, **7**(1-3), 16–39.

Feldman, David P. (2012). *Chaos and Fractals: An Elementary Introduction* (1st ed.). Oxford University Press.

Feldman, David P. (2014). Introduction to Dynamical Systems and Chaos. A Complexity Explorer Course: http://chaos.complexityexplorer.org.

Feldman, David P. (2015). Fractals and Scaling. A Complexity Explorer Course: http://fractals.complexityexplorer.org.

Flake, Gary W. (1999). *The Computational Beauty of Nature: Computer Explorations of Fractals, Chaos, Complex Systems, and Adaptation*. MIT Press.

Ford, Joseph (1983). How random is a coin toss? *Physics Today*, **36**(4), 40–47.

Frigg, Roman, Berkovitz, Joseph, and Kronz, Fred (2014). The ergodic hierarchy. In *The Stanford Encyclopedia of Philosophy* (Summer 2014 ed.) (ed. E. N. Zalta).

Frigg, Roman and Hartmann, Stephan (2012). Models in science. In *The Stanford Encyclopedia of Philosophy* (Fall 2012 ed.) (ed. E. N. Zalta).

Garfinkel, Alan, Shevtsov, Jane, and Guo, Yina (2017). *Modeling Life: The Mathematics of Biological Systems*. Springer.

Gaspard, Pierre (2005). *Rössler systems*, pp. 808–811. Routledge.

Ghys, Étienne (2015). The butterfly effect. In *The Proceedings of the 12th International Congress on Mathematical Education*, pp. 19–39. Springer.

Gladwell, Malcolm (1996). The tipping point: Why is the city suddenly so much safer—Could it be that crime really is an epidemic? *The New Yorker*, **72**(14), 32–38.

Gladwell, Malcolm (2002). *The Tipping Point: How Little Things Can Make a Big Difference*. Back Bay Books.

Gleick, James (1987). *Chaos: Making a New Science*. Penguin Books.

Griffith, William C. (1961). Constitutional law: Equal protection: Racial discrimination and the role of the state. *Michigan Law Review*, 1054–1077.

Guckenheimer, John and Holmes, Philip J. (2013). *Nonlinear Oscillations, Dynamical Systems, and Bifurcations of Vector Fields*, Volume 42. Springer Science & Business Media.

Gunawardena, Jeremy (2014). Models in biology: 'Accurate descriptions of our pathetic thinking'. *BMC Biology*, **12**(1), 29+.

Healy, Kieran (2016). Fuck nuance. *Sociological Theory*, **35**(2), 118–127.

Hegger, Rainer, Kantz, Holger, and Schreiber, Thomas (1999). Practical implementation of nonlinear time series methods: The TISEAN package. *Chaos: An Interdisciplinary Journal of Nonlinear Science*, **9**(2), 413–435.

Hénon, Michel (1976). A two-dimensional mapping with a strange attractor. In *The Theory of Chaotic Attractors*, pp. 94–102. Springer.

Hilborn, Robert (2002). *Chaos and Nonlinear Dynamics: An Introduction for Scientists and Engineers: 2nd ed.* Oxford University Press.

Hilborn, Robert C. (2004). Sea gulls, butterflies, and grasshoppers: A brief history of the butterfly effect in nonlinear dynamics. *American Journal of Physics*, **72**(4), 425–427.

Hirsch, Morris W., Smale, Stephen, and Devaney, Robert (2004). *Differential Equations, Dynamical Systems, and an Introduction to Chaos* (2nd ed.). Academic Press.

Hobbs, Jesse (1991). Chaos and indeterminism. *Canadian Journal of Philosophy*, **21**(2), 141–164.

Hoefer, Carl (2016). Causal determinism. In *The Stanford Encyclopedia of Philosophy* (Spring 2016 ed.) (ed. E. N. Zalta). Metaphysics Research Lab, Stanford University.

Howard, Brian C. (May 2016). Pablo Escobar's escaped hippos are thriving in Colombia. *National Geographic*.

Hutter, M. (2007). Algorithmic information theory. *Scholarpedia*, **2**(3), 2519. Revision #186543.

Kantz, Holger and Schreiber, Thomas (2004). *Nonlinear Time Series Analysis* (2nd ed.). Cambridge University Press.

Kaper, Hans and Engler, Hans (2013). *Mathematics and Climate*. Society for Industrial & Applied Mathematics, U.S.

Kaplan, Daniel and Glass, Leon (1995). *Understanding Nonlinear Dynamics*. Springer-Verlag.

Keller, Evelyn F. (2005). Revisiting "scale-free" networks. *Bioessays*, **27**(10), 1060–1068.

Kellert, Stephen H. (1993). *In the Wake of Chaos: Unpredictable Order in Dynamical Systems*. University of Chicago Press.

Kolata, Gina B. (1977). Catastrophe theory: The emperor has no clothes. *Science*, **196**(4287), 287–351.

Kolmogorov, Andrei N. (1965). Three approaches to the quantitative definition of information. *Problems of Information Transmission*, **1**(1), 1–7.

Kraul, Chris (December 2006). A hippo critical situation. *Los Angeles Times*.

Kremer, William (June 2014). Pablo Escobar's hippos: A growing problem. *BBC World Service*.

Laplace, Pierre-Simon (1951). *A Philosophical Essay on Probabilities*. Dover Publications, New York.

Letellier, Christophe and Messager, Valérie (2010). Influences on Otto E. Rössler's earliest paper on chaos. *International Journal of Bifurcation and Chaos*, **20**(11), 3585–3616.

Letellier, Christophe and Rössler, Otto E. (2006). Rössler attractor. *Scholarpedia*, **1**, 1721+.

Levins, Richard (1966). The strategy of model building in population biology. *American Scientist*, **54**(4), 421–431.

Levins, Richard (2006). Strategies of abstraction. *Biology and Philosophy*, **21**(5), 741–755.

Libchaber, A., Laroche, C., and Fauve, S. (1982). Period doubling cascade in mercury, a quantitative measurement. *Journal de Physique Lettres*, **43**(7), 211–216.

Libchaber, A. and Maurer, J. (1982). A Rayleigh Bénard experiment: Helium in a small box. In *Nonlinear Phenomena at Phase Transitions and Instabilities*, pp. 259–286. Springer.

Lind, Douglas and Marcus, Brian (1995). *An Introduction to Symbolic Dynamics and Coding*. Cambridge University Press, New York.

Linsay, Paul S. (1981). Period doubling and chaotic behavior in a driven anharmonic oscillator. *Physical Review Letters*, **47**, 1349–1352.

Lorenz, Edward N. (1963). Deterministic nonperiodic flow. *Journal of Atmospheric Sciences*, **20**, 130–148.

Lorenz, Edward N. (1993). *The Essence of Chaos* (Reprint ed.). University of Washington Press.

Mangel, Marc (2006). *The Theoretical Biologist's Toolbox: Quantitative Methods for Ecology and Evolutionary Biology*. Cambridge University Press.

May, Robert M. (1976). Simple mathematical models with very complicated dynamics. *Nature*, **261**, 459–467.

May, Robert M. (2002). The best possible time to be alive. In *It Must be Beautiful: Great Equations of Modern Science* (ed. G. Farmelo), pp. 28–45. Granta Books.

May, Robert M. (2004). Uses and abuses of mathematics in biology. *Science*, **303**(5659), 790–793.

Michel, Jean-Baptiste, Shen, Yuan K., Aiden, Aviva P., Veres, Adrian, Gray, Matthew K., Team, The Google Books, Pickett, Joseph P., Hoiberg, Dale, Clancy, Dan, Norvig, Peter, Orwant, Jon, Pinker, Steven, Nowak, Martin A., and Aiden, Erez L. (2011). Quantitative analysis of culture using millions of digitized books. *Science*, **331**(6014), 176–182.

Miller, John H. (2016). *A Crude Look at the Whole: The Science of Complex Systems in Business, Life, and Society*. Basic Books.

Miller, John H. and Page, Scott E. (2007). *Complex Adaptive Systems: An Introduction to Computational Models of Social Life* (illustrated edition). Princeton University Press.

Mitchell, Melanie (2009). *Complexity: A Guided Tour*. Oxford University Press.

Mitchell, Melanie (2013). Introduction to Complexity. A Complexity Explorer Course: https://intro.complexityexplorer.org/.

Mitzenmacher, Michael (2004). A brief history of generative models for power law and lognormal distributions. *Internet Mathematics*, **1**(2), 226–251.

Mora, Thierry and Bialek, William (2011). Are biological systems poised at criticality? *Journal of Statistical Physics*, **144**(2), 268–302.

Mukerjee, Madhusree (1996). Seeing the world in a snowflake. *Scientific American*, **274**(3), 36–42.

Newman, Mark (2012). *Computational Physics*. CreateSpace Independent Publishing Platform.

Newman, M. E. J. (2005). Power laws, Pareto distributions and Zipf's law. *Contemporary Physics*, **46**, 323–351.

Newman, M. E. J. (2011). Resource letter CS1: Complex systems. *American Journal of Physics*, **79**(8), 800–810.

Nolte, David D. (2010). The tangled tale of phase space. *Physics Today*, **63**(4), 33+.

O'Connor, Timothy and Wong, Hong Y. (2015). Emergent properties. In *The Stanford Encyclopedia of Philosophy* (Summer 2015 ed.) (ed. E. N. Zalta). Metaphysics Research Lab, Stanford University.

Ott, Edward, Sauer, Tim, and Yorke, James A. (1994). *Coping with Chaos: Analysis of Chaotic Data and The Exploitation of Chaotic Systems* (1st ed.). Wiley-Interscience.

Otto, Sarah P. and Day, Troy (2007). *A Biologist's Guide to Mathematical Modeling in Ecology and Evolution* (illustrated edition). Princeton University Press.

Packard, N. H., Crutchfield, J. P., Farmer, J. D., and Shaw, R. S. (1980). Geometry from a time series. *Physical Review Letters*, **45**.

Palmer, T. N., Shutts, G. J., Hagedorn, R., Reyes, F. J. Doblas, Jung, T., and Leutbecher, M. (2005). Representing model uncertainty in weather and climate prediction. *Annual Review of Earth and Planetary Sciences*, **33**(1), 163–193.

Pecora, Louis M., Moniz, Linda, Nichols, Jonathan, and Carroll, Thomas L. (2007). A unified approach to attractor reconstruction. *Chaos: An Interdisciplinary Journal of Nonlinear Science*, **17**(1), 013110.

Peitgen, Heinz-Otto, Jürgens, Hartmut, and Saupe, Dietmar (1992). *Chaos and Fractals: New Frontiers of Science*. Springer-Verlag.

Peterson, Roger T. (1989). *A Field Guide to the Birds: Eastern and Central North America* (illustrated, large print ed.). Houghton Mifflin Harcourt.

Poston, Tim and Stewart, Ian (2012). *Catastrophe Theory and Its Applications*. Dover Publications.

Press, W. H., Teukolsky, S. A., Vetterling, W. T., and Flannery, B. P. (1995). *Numerical Recipes in C: The Art of Scientific Computing* (2nd ed.). Cambridge University Press.

Railsback, Steven F. and Grimm, Volker (2011). *Agent-Based and Individual-Based Modeling: A Practical Introduction*. Princeton University Press.

Reed, William J. and Hughes, Barry D. (2002). From gene families and genera to incomes and internet file sizes: Why power laws are so common in nature. *Physical Review E*, **66**(6), 067103+.

Robinson, R. Clark (2012). *An Introduction to Dynamical Systems: Continuous and Discrete (Pure and Applied Undergraduate Texts)* (2nd ed.). American Mathematical Society.

Ross, Chip, Odell, Meredith, and Cremer, Sarah (2009). The shadow-curves of the orbit diagram permeate the bifurcation diagram, too. *International Journal of Bifurcation and Chaos*, **19**(09), 3017–3031.

Ross, Chip and Sorensen, Jody (2000). Will the real bifurcation diagram please stand up! *The College Mathematics Journal*, **31**(1), 2.

Rössler, Otto E. (1976). An equation for continuous chaos. *Physics Letters A*, **57**(5), 397–398.

Ruelle, David (1980). Strange attractors. *The Mathematical Intelligencer*, **2**(3), 126–137.

Ruelle, David (1993). *Chance and Chaos*. Princeton University Press.

Ruelle, David and Takens, Floris (1971). On the nature of turbulence. *Communications in Mathematical Physics*, **20**(3), 167–192.

Sauer, Tim, Yorke, James A., and Casdagli, Martin (1991). Embedology. *Journal of Statistical Physics*, **65**(3-4), 579–616.

Sauer, Timothy D. (2006). Attractor reconstruction. *Scholarpedia*, **1**(10), 1727+.

Sayre, Nathan F. (2008). The genesis, history, and limits of carrying capacity. *Annals of the Association of American Geographers*, **98**(1), 120–134.

Scheffer, Marten (2009). *Critical Transitions in Nature and Society*. Princeton University Press.

Scheffer, Marten, Carpenter, Stephen R., Lenton, Timothy M., Bascompte, Jordi, Brock, William, Dakos, Vasilis, Van De Koppel, Johan, Van De Leemput, Ingrid A., Levin, Simon A., Van Nes, Egbert H., and Others (2012). Anticipating critical transitions. *Science*, **338**(6105), 344–348.

Schuster, Heinz G. and Just, Wolfram (2006). *Deterministic Chaos: An Introduction*. John Wiley & Sons.

Servedio, Maria R., Brandvain, Yaniv, Dhole, Sumit, Fitzpatrick, Courtney L., Goldberg, Emma E., Stern, Caitlin A., Van Cleve, Jeremy, and Yeh, D. Justin (December 2014). Not just a theory—The utility of mathematical models in evolutionary biology. *PLoS Biol*, **12**(12), e1002017+.

Smith, Leonard (2007). *Chaos: A Very Short Introduction*. Oxford University Press.

Smith, Peter (1998). *Explaining Chaos*. Cambridge University Press.

Solomonoff, Ray J. (1964a). A formal theory of inductive inference. Part I. *Information and Control*, **7**(1), 1–22.

Solomonoff, Ray J. (1964*b*). A formal theory of inductive inference. Part II. *Information and Control*, 7(2), 224–254.

Sprott, Julien C. (2003). *Chaos and Time-Series Analysis*. Oxford University Press.

Stein, Daniel L. and Newman, Charles M. (2013). *Spin Glasses and Complexity*. Princeton University Press.

Stewart, Ian (2002). *Does God Play Dice? The New Mathematics of Chaos* (2nd ed.). Wiley-Blackwell.

Strogatz, Steven (2001). *Nonlinear Dynamics and Chaos: With Applications to Physics, Biology, Chemistry and Engineering*. Perseus Books.

Stumpf, Michael P. H. and Porter, Mason A. (2012). Critical truths about power laws. *Science*, **335**(6069), 665–666.

Takens, Floris (1981). Detecting strange attractors in turbulence. *Lecture Notes in Mathematics*, **898**(1), 366–381.

Takens, Floris (2010). Reconstruction theory and nonlinear time series analysis. In *Handbook of Dynamical Systems* (B. Hasselblatt, H. W. Broer, and F. Takens eds.), Volume 3, 345–377. Elsevier.

Terman, D. H. and Izhikevich, E. M. (2008). State space. *Scholarpedia*, **3**(3), 1924. revision #91820.

Thom, René (1994). *Structural Stability And Morphogenesis (Advanced Books Classics)*. Westview Press.

Thompson, Clive (2008). Is the Tipping Point Toast? *Fast Company*. http://www.fastcompany.com/641124/tipping-point-toast, accessed February 5, 2015.

Thompson, Silvanus P. and Gardner, Martin (1998). *Calculus Made Easy* (revised, updated, expanded ed.). St. Martin's Press.

Tresser, C., Coullet, P., and Faria, E. De (2014). Period doubling. *Scholarpedia*, **9**(6), 3958. Revision #142883.

Vandermeer, John H. and Goldberg, Deborah E. (2013). *Population Ecology: First Principles* (2nd ed.). Princeton University Press.

Watts, Duncan J. (2004). *Six Degrees: The Science of a Connected Age* (Reprint ed.). W. W. Norton & Company.

Watts, Duncan J. (2012). *Everything Is Obvious: How Common Sense Fails Us*. Crown Business.

Watts, Duncan J. and Dodds, Peter S. (2007). Influentials, networks, and public opinion formation. *Journal of Consumer Research*, **34**(4), 441–458.

West, Geoffrey (2017). *Scale: The Universal Laws of Growth, Innovation, Sustainability, and the Pace of Life in Organisms, Cities, Economies, and Companies*. Penguin.

West, Geoffrey B. and Brown, James H. (2004). Life's universal scaling laws. *Physics Today*, **57**(9), 36–43.

West, Geoffrey B., Brown, James H., and Enquist, Brian J. (1997). A general model for the origin of allometric scaling laws in biology. *Science*, **276**(5309), 122–126.

Wikipedia (2016). Lotka–Volterra equations—Wikipedia, the free encyclopedia. [Online; accessed August 23, 2016].

Wilensky, Uri and Rand, William (2015). *An Introduction to Agent-Based Modeling: Modeling Natural, Social, and Engineered Complex Systems with NetLogo*. The MIT Press.

Woodcock, Alexander and Davis, Monte (1978). *Catastrophe Theory*. Dutton.

Yeomans, J. M. (1992). *Statistical Mechanics of Phase Transitions*. Clarendon Press, Oxford.

Zahler, Raphael S. and Sussmann, Hector J. (1977). Claims and accomplishments of applied catastrophe theory. *Nature*, **269**(5631), 759–763.

Zeeman, E. C. (1976). Catastrophe theory. *Scientific American*, **234**(4), 65–83.

Zeeman, E. C. (1977). *Catastrophe Theory: Selected Papers, 1972-77*. Addison-Wesley Educational Publishers Inc.

INDEX